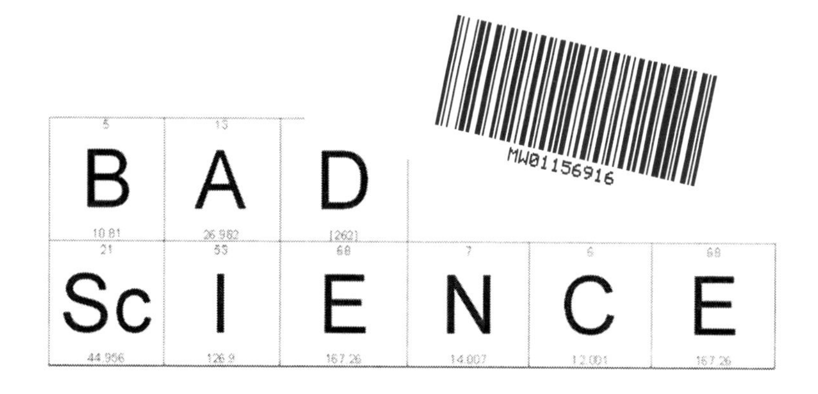

BAD SCIENCE

A Brief History of Bizarre Misconceptions, Totally Wrong Conclusions and Incredibly Stupid Theories

LiNDA ZIMMERMANN

Eagle Press
New York

For more information, or to contact the author, go to:

www.badsciences.com

Or write to:

Linda Zimmermann
P.O. Box 192
Blooming Grove, NY 10914

Bad Science
Copyright © 2011 Linda Zimmermann

All rights reserved. This book may not be reproduced in whole or in part without permission.

ISBN: 978-0-9799002-4-2

CONTENTS

Author's Note

I love science.

Although I didn't realize it at the time, in retrospect I freely admit I was a science geek when I was a kid. I charted sunspots, collected bugs, built models of spacecraft, mixed household chemicals and cleaning products to see how they would react, took things apart to see how they worked, and enthusiastically tried to learn everything I could.

Even though I also loved to write and knew someday I would give it a shot, there was never any question that I would first pursue a career in science. While in college, I got a part-time job working in the Quality Control microbiology lab of a medical diagnostics company. I moved over to chemistry QC, and after graduation, became a full-time employee in the Research and Development department.

I wore the requisite white lab coat, the nerdy safety glasses and safety shoes, and was completely enamored of all the glassware, chemicals, and instrumentation. What I didn't like was the company politics, the sales and marketing people who were treated like demigods (while the scientists who created the products they sold were clearly second-class citizens), and the arrogance and outright dishonesty of some of the scientists who felt that higher degrees were something akin to being members of the aristocracy.

The writer in me stirred. These people were sullying the purity of science, and I became rather miffed. In response, I wrote a satirical newsletter called the *Narwhal Gazette* (it's a long story), and lampooned the company's people, policies, and projects. To my astonishment, I didn't get fired! In fact, everyone—including the bosses and stuffy scientists—loved it, and people began lining up at the copier to get the latest issue "hot off the presses." People actually *wanted* to be written about, and I was emboldened to be even more outrageous and daring in my satire.

The *Gazette* flourished for many years, until new management came in. They were not amused. I had two options—stop writing the *Gazette,* or continue with the agreement that all my articles would be approved and edited by management. Inflamed with righteous indignation, there was no question that I would sooner stand before a firing squad, before I would submit to state-sanctioned censorship, so the farewell issue of the *Gazette* signaled the end to my humorous barbs aimed at the world of science.

Or so I thought.

When the company eliminated the R&D department, I didn't seek out another job in a lab. The magnetic pull of writing took hold and I began working on short stories, novels, history articles and books, and articles on astronomy. I also started lecturing on astronomy, and came upon the idea for a humorous program about the history of all the crazy things that were once believed. One thing led to another, and in 1995, I published the book *Bad Astronomy: A Brief History of Bizarre Misconceptions, Totally Wrong Conclusions, and Incredibly Stupid Theories*.

I gave presentations on *Bad Astronomy* at "Star Parties" and astronomy conventions from New England to Florida, and enjoyed every minute of it. There was the occasional audience member who thought I was being too critical of those throughout history who had committed acts of Bad Astronomy, but the overwhelming majority of people just laughed and had a good time.

I went on to many other writing projects over the years, but all the while kept my eyes open for similar examples in other fields of science, and added those stories to a folder I marked "Bad Science." I knew some day another book would emerge from that overstuffed folder, and that day came in 2010, when the magnetic pull of science—and memories of the *Gazette*—signaled that the time had come.

As much as I enjoy writing on a wide variety of topics, plunging headlong into *Bad Science* was a wonderful revelation. This is who I was—the geeky science kid, grown up into the wisecracking author/lecturer, with an irresistible urge to smite the foes of Good Science. It was a project I never wanted to end, but as sad as I was to complete the manuscript, I was overjoyed at the prospect of sharing my irreverent views with readers and audiences again.

So, where do I stand on what constitutes Bad Science? For starters, the short answer would be torturing someone like Galileo for his heliocentric theories. Then there are those who commit fraud or harm people because of greed and ego. There's also the ever-popular ignoring the obvious to perpetuate one's own agenda, and refusing to evaluate the merits of the facts because of personal and religious beliefs.

On this last point, I need to make special mention. When it comes to personal beliefs, I am fully aware that my way of thinking isn't exactly in the mainstream, and I believe in some things that others would consider completely absurd, weird, and unscientific. But at least I *know* they are unscientific, and always try to maintain a separation between personal beliefs and science.

But enough of all this. The purpose of this book is to amuse you, the reader, and if you learn something along the way, so much the better. Judge for yourself the merits of each case presented here. And if you happen to work in a lab, hospital, or research facility, keep your eyes and ears open—for Bad Science is always lurking in the shadows, waiting to strike!

Linda Zimmermann
January 2011

Medicine

A Shocking Experience

In 1781, Luigi Galvani made a dead frog jump by applying an electric current to its muscles. Immediately, dissatisfied young housewives dreamed of using this technique on their aging husbands, while physicians envisioned a greater benefit, i.e., to their bank accounts. Operating under the broad assumption that hefty doses of electricity had a stimulating and therapeutic effect on people, patients everywhere were soon dishing out the cash to be zapped by various devices. Most of these devices impressed patients with their noise and the tingling sensations and pain they produced, but the scientific community eventually realized they had no real medical value.

While the use of properly controlled electricity can be of enormous medical value today (for instance, the "paddles" used to get a heart beating again), the early "medical batteries" can be elevated no higher than to the level of amusing parlor tricks. However, if a study of medicine has proven anything, it is that some people never give up.

Enter Charles Willie Kent in 1918, who fancied himself the living embodiment of both Euclid, the ancient Greek mathematician, and Abraham Lincoln. His claim to fame was the *Electreat Mechanical Heart*, which, curiously enough, was neither mechanical nor any type of pumping apparatus. It was, in fact, a throwback to the old electric devices of the previous century, with a rather disturbing twist—thin metal rods which could be inserted into the orifice of your choice.

Beyond providing nasty shocks in unspeakable places, the *Electreat* was supposed to stop pains of all sorts, and cure everything from dandruff to glaucoma to appendicitis. Kent even claimed that his device had the power to enlarge women's breasts. All this for the low price of $15 retail, and at $7 wholesale, the 229,273 units Kent sold made him a tidy little fortune.

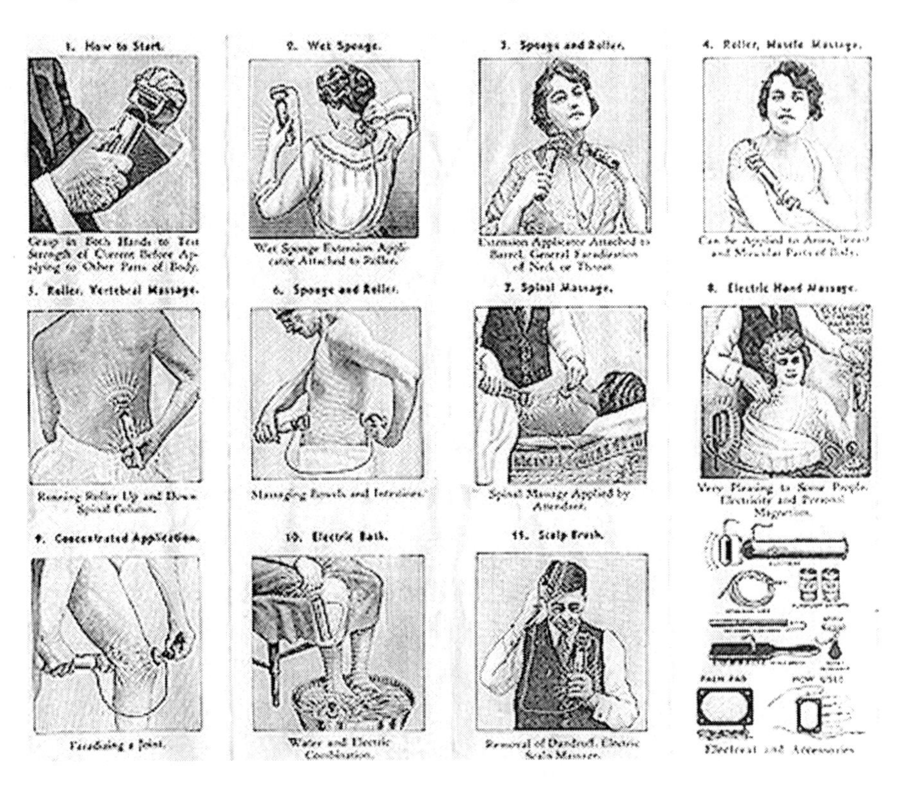

The *Electreat's* many useless uses.

There was only one small problem. The only thing the *Electreat* helped was Kent's cash flow. In 1941, the FDA finally brought Kent to court to answer to his wild claims of the device's miraculous powers. The government paraded an impressive line of experts who systematically demolished Kent's assertions.

For example, the primary benefit of the *Electreat* was supposed to be increased blood circulation. A biologist from the University of

Kansas explained in detail how the device actually caused the opposite reaction, effectively stopping the flow of blood in the unfortunate muscle to which the *Electreat's* current was being applied (information which no doubt brought great disappointment to those dissatisfied housewives).

After the blistering testimony for the prosecution, Kent, against the better judgment of his lawyer, decided to take the stand. Repeatedly embarrassing himself with his medical and electrical ignorance, his crowning blunder came when asked if he used the *Electreat* on his own body.

"Yes, sir," Mr. Kent boldly replied, "For menopause!"

With the court's blessing, Kent's stock of *Electreat Mechanical Hearts* was rounded up and destroyed. Defeated, but undaunted, Kent was determined to resume production. Shortages due to World War II kept him from his goal until 1946, when *Electreats* once again appeared on the market, albeit with greatly subdued claims. Once again, Kent and the experts met in court, and once again the government ably proved its point.

This time, Kent could have gone to prison for endangering and duping a gullible populace, but due to his advanced age, he was fined a mere $1,000. With his dreams and business shattered, perhaps he was able to find some little comfort in his twilight years with his fortune, and a well-placed metal rod.

"Fooling around with alternating current is just a waste of time. Nobody will use it, ever."
Thomas Edison, 1889

Animal Magnetism

Today, animal magnetism alludes to one's sexual allure, but in the 1700s it referred to a mysterious force that was supposed to have the power to heal. Promoted by Dr. Franz Anton Mesmer, animal magnetism may go down in history as one of the greatest marketing ploys, as well as one of the greatest frauds.

For starters, Dr. Mesmer earned his degree the old fashioned way—plagiarism. The dissertation from which he "borrowed"

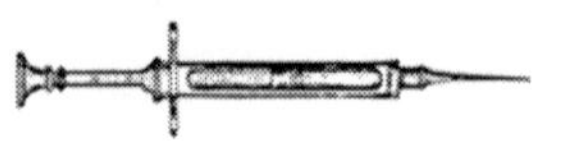

concerned the influence of the planets on health, so it was no great leap to his contention that there was a "universal magnetic fluid," and *he* had the ability to control it with various devices.

However, even without these devices, Mesmer did appear to have quite the magnetic personality. He developed a form of hypnosis that worked rather effectively with his patients—especially on the impressionable "hysterical bourgeois women" that seemed to predominate his clientele. Mesmer was said to have been able to make people fall into a trance, have convulsions, dance, and best of all, think they were cured. His prowess is forever immortalized in the popularly used term "mesmerized," which today still describes a hypnotic or spellbinding effect.

But let us return to animal magnetism and the creative ways Mesmer used to exploit it. First, it should be explained where he got the idea to use magnetism—he stole it. This time, it was from a Jesuit named Maximilian Hell. The regrettably named Father Hell used a steel plate that was supposedly magnetized to heal people. Many claimed that the metal plate cured what ailed them, so Mesmer took the concept and ran with it—eventually running all the way to Paris where he was to reach the heights of the royal court, and the pinnacle of quack medicine.

As Mesmer described the principles of his healing powers, the human body often suffered from blockages and disruptions to the flow of magnetic fluid, and his therapies corrected that flow. This was achieved by touching the patient with metal rods that were usually connected to a fluid that allegedly created electricity and magnetic power. He even came up with a tub of fluid in which the patient sat and was then touched with metal rods. However, when crowds began to gather for treatments, Mesmer realized he needed to expand and create mass magnetic devices.

Actually asserting that he could magnetize just about anything, Mesmer claimed to have magnetized a large tree. He then hung ropes out of this magnetized tree, and patients would hold on to the ropes to receive their remarkable treatments. Sticking with the original tub concept, he also constructed "baquets." These were big wooden tubs about five feet in diameter, which contained various bottles and iron filings submerged in a foot of water, the whole of which was supposed to act something like a battery for magnetic fluid. Metal rods were placed in the walls of the tub, and patients would rest their afflicted regions against the rods.

Mesmer's greatest performances, however, must have been those he conducted while wearing colorful robes, like Merlin working his magic on wealthy and clueless audiences. Combining hypnosis, his magnetic tubs, and a level of showmanship that would have brought a tear to eye of P.T. Barnum, he became the darling of the court of Louis XVI. Until the party ended in 1784…

A group of scientists—including Ben Franklin, who knew a thing or two about electricity—examined Mesmer's tubs and devices, and found nothing. No electricity, no powerful mysterious forces, no trace of animal magnetism. Finally exposed as a fraud, Mesmer's therapies were officially banned.

Mesmer eventually left Paris and lived a quiet life until his death at the age of 85, which wasn't too shabby for those days. Perhaps he did know something after all?

Unfortunately, the magic of magnetism still has the power to separate naïve people from their money. Magnetic bracelets, bands, belts, and even self-adhesive magnets to stick wherever you like, are still marketed in catalogs, infomercials, and alternative healing clinics. In retrospect, the success of these items may speak more to the failure of modern medicine to cure what ails us—or at least be able to do it at a price the average person can afford!

So go ahead, toss some ropes into a tree, build a baquet in your backyard, throw in some iron filings, add a few metal rods, and invite your friends over for a mesmerizing evening…

Where's your Sense of Humour?

For thousands of years, the practice of medicine was based upon the four humours, or fluids, that supposedly determined a person's

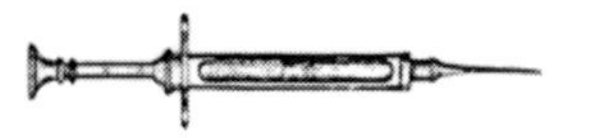

temperament and physical characteristics, and they also led to disease if they were out of balance. These humours were yellow bile, black bile, blood, and phlegm, or as Hippocrates referred to them—cholera, melankholia, sanguis, and phlegma.

If the humours were in equilibrium, a person would be healthy and have an even temperament. However, if a person was thought to have too much blood, for instance, he was prone to one of the "hot" diseases and fever, and would need to be bled to restore order. Of course, too much bleeding would lead to death, but at least the corpse would have a balanced sense of humour.

It was also a commonly held belief that many things could adversely affect the humours, such as the seasons, mysterious "vapors" in the air, as well as the food one ate. The following chart shows some of the alleged relationships with the various humours.

Humour	Season	Element	Organ	Qualities	Temperament
Sanguine/ Blood	spring	air	liver	Hot/Wet	bravery, love
Choleric/ Yellow Bile	summer	fire	gall bladder	Hot/Dry	anger
Melancholic/ Black Bile	autumn	earth	spleen	Cold/Dry	depression
Phlegmatic/ Phlegm	winter	water	*brain/ lungs	Cold/Wet	calm

(* I think I know several people who have phlegm for brains.)

In all fairness, considering the fact that previous to the concept of humours, the practice of medicine was often nothing more than ceremonies to drive away evil spirits, this was all quite a big step forward. It at least got people thinking about the relationship of diet and exercise to overall health, and to start categorizing symptoms of various diseases and how they affect the organs.

The problem is, no real advancements beyond the belief in humours took place for many centuries. The erroneous and often harmful practices and treatments which attempted to balance the humours continued into the 19th century, and are actually still part of some modern systems of medicine. These ancient beliefs have also left their mark in our daily lives, as many terms that harken back to "humourism" have become permanent fixtures in our language, such as sanguine, melancholy, a "dry" wine, and "hot" spices.

It is impossible to imagine how many lives were lost over the millennia because of the belief in humours, but the road to Good Science is all too often paved with human suffering. Perhaps some people were actually helped by a more sensible diet, some exercise, and the judicious use of herbs. And perhaps, if there was a lot less yellow and black bile in the world today, mankind's humour would improve.

The Royal Touch

King Edward the Confessor (1002-1066) of England earned his name through his piety. However, as compassionate and devout as he was, it may have gone to his head just a bit.

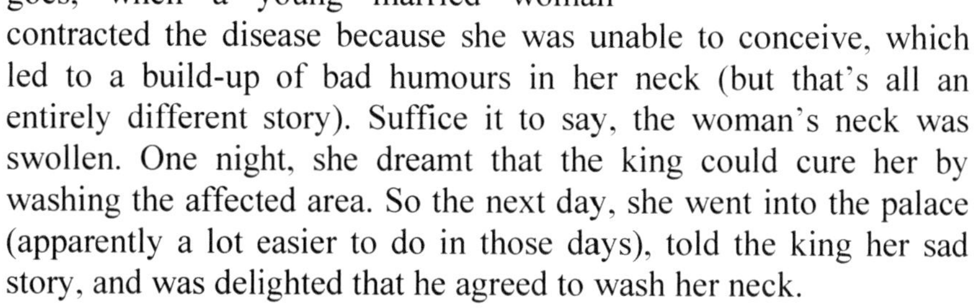

There was, at the time, a nasty disease known as scrofula; a swelling of the neck due to tuberculosis in the glands. The condition came to be known as the "king's evil," because King Edward believed he could cure it, starting a long tradition of a monarch's alleged ability to heal by touch.

The whole affair began, as the story goes, when a young married woman contracted the disease because she was unable to conceive, which led to a build-up of bad humours in her neck (but that's all an entirely different story). Suffice it to say, the woman's neck was swollen. One night, she dreamt that the king could cure her by washing the affected area. So the next day, she went into the palace (apparently a lot easier to do in those days), told the king her sad story, and was delighted that he agreed to wash her neck.

As good King Edward rubbed the swollen area, the skin ruptured and all manner of putrid fluids gushed out, accompanied by a substantial population of squirming, writhing worms (clearly not a typical case of tuberculosis). Barely a week later, the worms were all gone, the gaping wound had closed with no sign of scarring, and a faith healer was born. King Edward soon expanded his repertoire to

curing epilepsy and all types of ailments, although scrofula did remain his sentimental favorite.

Most amazing of all, the practice of the "Royal Touch" continued in France and England until the 18th century. Elaborate ceremonies were developed for the regal dispensation of the touch, and during a single ceremony, as many as 1,500 people would have the magic hand of the benevolent ruler placed upon their brows. In fact, it seems that just about everyone in those days was touched in the head.

The Manner of His Majesties Curing the Difeafe,
CALLED THE
KINGS-EVIL.

A broadside about the "healing" event.

However, not all monarchs believed themselves to be healers. Yet those who tried to stop the practice were vehemently criticized and accused of being cruel and heartless. King William III of England (1650-1702), himself a reluctant healer, dispensed something with each touch which was indeed valuable—a piece of wise advice. William would lay his hands upon the patient and utter the words, "May God give you better health and *more sense*."

Getting Your Head Read

A person's character and behavior are complex things, and can be influenced by environment, stress, diet, disease, and of course, hormones. Oh, and did I mention hormones?

Anyway, in the 1790s, a doctor in Vienna, Franz Gall, believed he discovered the "one true science of the mind," where the contours of the skull were signposts to an individual's nature. He drew these conclusions based upon the premise that the brain had separate and distinct "organs" for each of the various characteristics. The larger the organ, the more powerful that characteristic, and the more prominent that region of the skull would be. Similarly, a depressed region of the skull would indicate a lack of that particular tendency.

For example, there is a spot behind the left ear that supposedly corresponds with a person's innate courage. There are other areas for reasoning ability, dexterity, and perseverance. Then things get a little more bizarre as one explores the cranial peaks and valleys that denote "Hope for the Future," "Parental Love," "Spiritual Faith," "Worship," "Mirthfulness," "Desire for Liquids," "Patriotism," and "Philanthropy." Gall also believed there were brain organs and bumps for murder and theft. Some employers even required that job applicants have their heads examined by a specialist to ensure that they were honest, hardworking people.

While Gall called his discovery "organology," it became widely known as phrenology, meaning study of the mind. Phrenology was very popular in England in the 1820s-40s, and spread throughout the United States. Many texts were written, charts were drawn, and people paid good money to have their heads read by expert phrenologists.

There was just one little problem—it was all nonsense. Critics, of which there were many, termed the pseudoscience "Bumpology," and pointed out its many flaws. For example, if an individual had a prominent bump on the spot that indicated fidelity in marriage, but constantly cheated on his wife, phrenologists explained it away by claiming other organs of the brain had negatively influenced that part of the marriage organ.

But wait, if size didn't matter after all, didn't that completely negate the very foundation of phrenology? In fact, it did, but no one bothered to mention it. Phrenology continued to thrive and continued to attract suckers, including some very prominent suckers.

"I never knew I had an inventive talent until Phrenology told me so. I was a stranger to myself until then!" said Thomas Edison, perhaps the greatest inventor of all time. (Perhaps it's a good thing

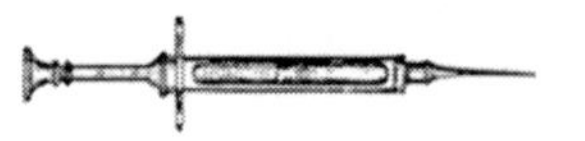

someone felt the bumps on his head or we might all still be sitting in the dark.)

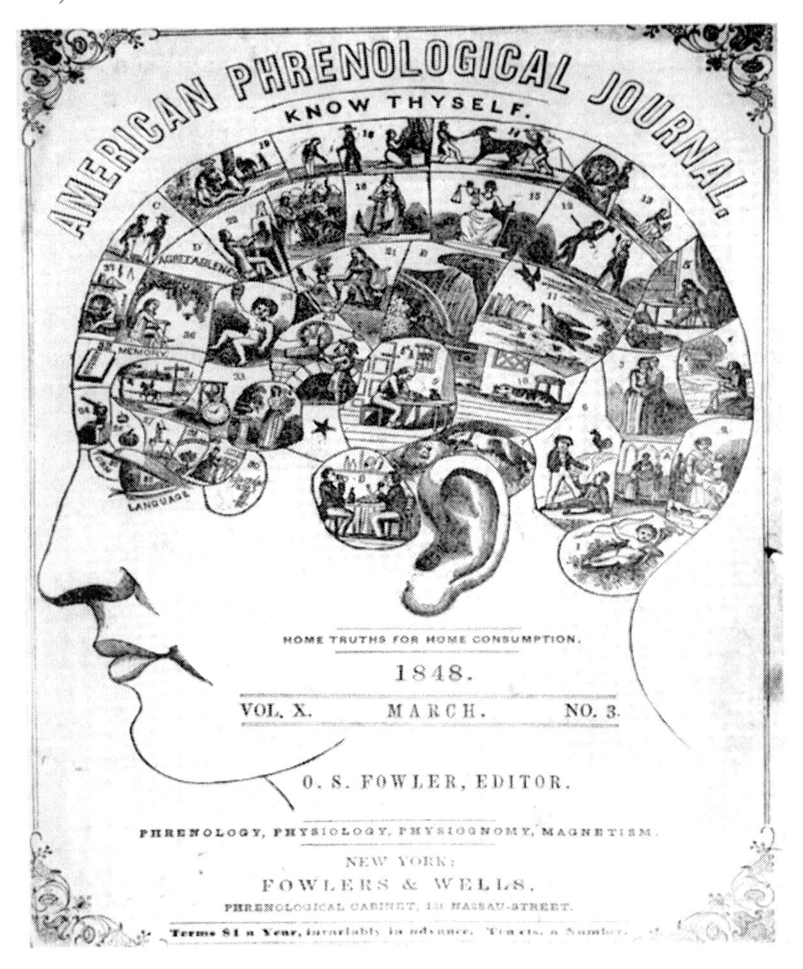

One of the primary reasons that phrenology continued to prosper for so long is that there was a lot of money to be made "reading" the skulls of the gullible. Even as late as 1934, visitors to the Century of Progress Exposition in Chicago shelled out a total of over $200,000 to have their coconuts analyzed by the "psychograph," a semi-circular cage of measuring devices that fit over the head and printed out a complete laundry list of your virtues and vices.

Today, one only needs to search the Internet to see that phrenology is still alive and kicking. One site that staunchly defends it, even plainly states that scientific criticisms are nothing but "crap." The author of that statement might want to check the "Combativeness" bump in the "Defense" region of his skull, not to mention the various peaks and valleys of stupidity, common sense, and the ability to extract one's head out of one's rear end. (Okay, I made up some of those, but it did make my "Wit" bump tingle...)

In all fairness, Gall and his subsequent followers did get some things right. There are indeed localized functions in the brain, and areas that are used extensively can grow larger. However, none of that has anything to do with your pate's protuberances.

The road to knowledge and truth has always been rocky, but in the unfortunate case of phrenology, there's no reason to make that road any bumpier.

> "I am tired of all this sort of thing called science here... We have spent millions in that sort of thing for the last few years, and it is time it should be stopped."
>
> Simon Cameron, U.S. Senator, speaking about the Smithsonian Institution, 1901

William Harvey

Many of the things we take for granted today were once taboo, and many of our most basic concepts were considered to be dangerous subjects that could cost you your career, or your life. Even though we have seen this resistance to new ideas century after century, it seems absurd to think that something as fundamental as the beat of a heart could have stirred up such a storm of controversy.

The first half of the 1600s wasn't the safest time in history to propose original ideas. Men of science in Catholic Europe were being threatened, tortured, and burned at the stake for trying to disrupt the old school of thought with new theories based upon observations and experiments. In other words, they were in peril simply because they were being good scientists.

So it was with some trepidation, and twelve long years of deliberation, that Englishman William Harvey finally published his theories on the heart and blood. In 1628, he presented the little 72-

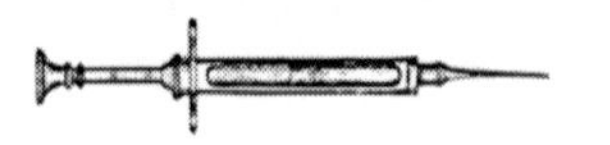

page book with the big name: *Exercitatio Anatomica de Motu Cordis et Sanguinis in Animalibus (An Anatomical Exercise on the Motion of the Heart and Blood in Living Beings).* (To cut down on fatigue we shall henceforth refer to it as "Harvey's book.")

And what was in Harvey's book that caused such a ruckus? Here's the monumental, earth-shattering revelation: The heart pumps the blood that circulates through the body.

No, really, that's it. Harvey said that the heart is a pump and blood circulates through the veins and arteries. That was it in a nutshell. Doesn't sound like anything to get upset about, right?

Wrong! For over twenty years after Harvey's book was published, he endured scathing criticism, with some "learned" men going so far as to declare that his theories violated the very laws of God and nature.

No, really. People were quite upset by the idea of the heart pumping circulating blood.

The reason they were so upset was that over 1400 years earlier the legendary Greek physician Galen had written that blood was created in the liver, went one way through the veins out to the edges of the body where it just disappeared. According to his theory, blood was essentially disposable—use it once and throw it away. And the heart was not a pump, it was the seat of a spiritual force. And arteries were not meant to carry blood, they were designed to transport air and this mystical spirit.

Galen became so revered over the centuries that his work was taken as gospel. And anyone who challenged Galen was a fool and a heretic. If you conducted experiments that produced evidence that went against Galen, it could only mean one thing—*your* experiment was faulty and *your* conclusions were wrong. (*And* you were a fool and a heretic.)

Why people who are supposedly educated have such a hard time thinking for themselves I'll never know, but poor William Harvey courageously continued to try to demonstrate that he was right. Of course, he was missing one very big piece of the puzzle—or one very tiny, microscopic piece, to be more exact—capillaries, those smallest of vessels that act as conduits between the arteries and veins. Even though Harvey was certain that blood circulated, he couldn't explain how it bridged that gap.

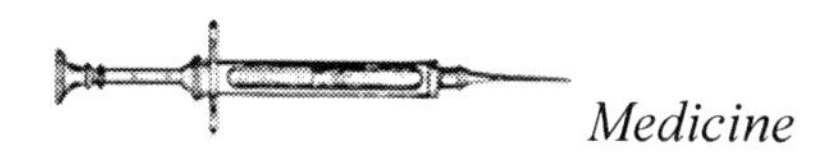

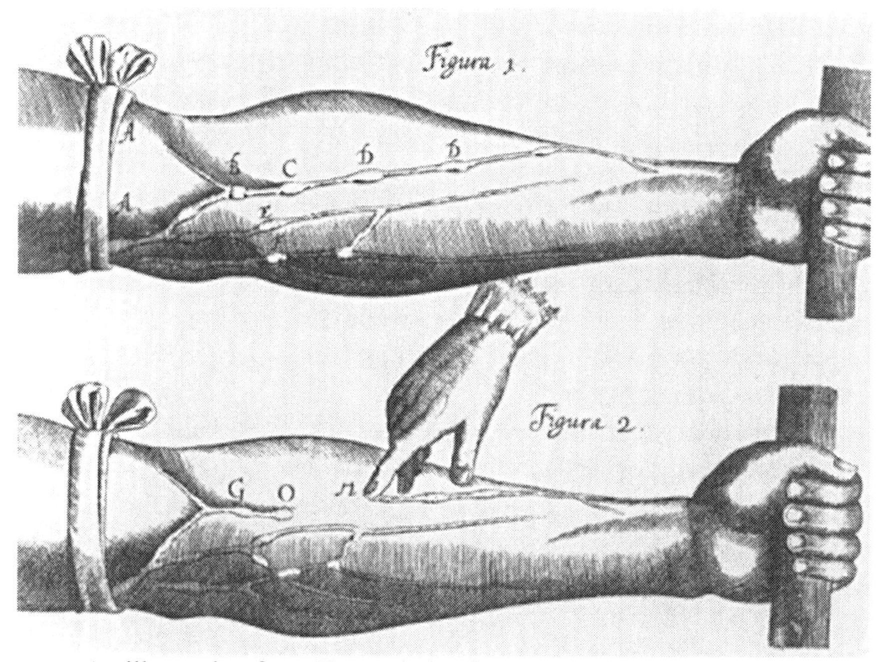

An illustration from Harvey's book. Note the importance of the researcher's wrist having the proper lace cuff.

Yet, there was plenty of other evidence, including a very simple calculation. By measuring the capacity of the left ventricle—the chamber that was the last stop for blood in the heart before it headed out into the body—and multiplying that number by how many times the heart beats, Harvey determined that if new blood was constantly being produced as Galenists believed, the body would conservatively have to make 65 quarts of blood every hour! That's 1,560 quarts a day, or 569,400 quarts every year. Critics countered by asserting that since the heart *wasn't* a pump, Harvey's calculations were meaningless.

If your blood isn't boiling by this point, consider this—Harvey was still taking flak over twenty years after the publication of his book, even though many physicians had finally understood the truth. In a historic show of restraint, Harvey waited until 1649 to respond in writing to his steadfastly ignorant critics. Fortunately, he was comfortably confident in his results and had attained enough

recognition for his achievements that he was able to rise above those critics and feel satisfied with his life's work.

Still, William Harvey died in 1657 not knowing the missing piece to his circulatory puzzle. He had survived to the ripe old age of 79, but had he made it to 83, he would have had his answer—and complete vindication.

In 1661, Italian physician Marcello Malpighi was studying the lungs of a frog with a microscope, and was the first man to observe capillaries. Here was the missing link, the fine network of tiny vessels that cemented the connection between the heart, arteries, veins, and the considerably less than 65 quarts of blood that circulated through our bodies.

It's a shame that Harvey didn't live to see through Malpighi's microscope, but it warms my heart to imagine that somehow Harvey was looking over his colleague's shoulder at that moment of discovery…

Spanish physician Michael Servetus (born 1511) was a highly educated man who also studied astronomy, meteorology, geography, anatomy, and pharmacology. Servetus might have been credited with the discovery of circulation, had his books been more widely known. The problem was, he was considered a heretic because of his anti-trinity beliefs, so Servetus and all but three copies of his books were burned in 1553. (In a thoughtful gesture by his executioners in Geneva, one of Servetus' books was chained to his leg when he was burned at the stake.)

"The abdomen, the chest, and the brain will forever be shut from the intrusion of the wise and humane surgeon."

Sir John Eric Ericksen, Surgeon Extraordinary to Queen Victoria, 1873

"The abolishment of pain in surgery is a chimera. It is absurd to go on seeking it...knife and pain are two words in surgery that must forever be associated in the consciousness of the patient."

Dr. Alfred Velpeau, French surgeon, 1839

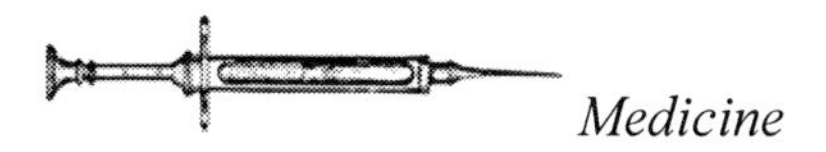

Keep them From Harm...

For centuries, physicians have sworn the Hippocratic Oath. The oath speaks to the nobility of the profession and the integrity under which one must conduct oneself at all times. And it clearly states that a doctor's primary obligation to his patients is to "keep them from harm and injustice." While it can be argued that everything is open to some interpretation, in this case, murder might be stretching the envelope a wee bit too far.

In the early 1800s, students flocked to the medical school in Edinburgh, Scotland. Anatomy was all the rage, and one of the foremost anatomists was Dr. Robert Knox. It was not uncommon for Knox to have five hundred students per class, and each potential new doctor was thrilled to explore the secrets of the human body with such a renowned expert. Unfortunately, Dr. Knox had quite a few secrets where bodies were concerned...

At the time, there were severe restrictions regarding dissections of human cadavers. The law stated that a school was only permitted to have one cadaver per year, and it had to be the body of a criminal who had been executed. Five hundred students in just one class, and only one corpse to go around! Naturally, this situation was quite frustrating to Knox and the other doctors trying to teach valuable lessons that could save lives. So, they decided to start a little "underground" trade in dead bodies.

It seems there were certain people in the city who thought it was a great waste to leave perfectly good corpses in their graves. They realized there was a veritable post-mortem gold mine beneath the soil of Edinburgh's cemeteries, so they started a profitable mining operation—although some more narrow-minded citizens viewed it as grave robbing.

Always on the lookout for recent burials, these men skulked about cemeteries in the dead of night, exhuming the bodies of those who had just been put to rest in eternal peace—giving them a slight detour to a slab in a medical school to be cut up into little pieces beneath the curious eyes of hundreds of students. These "resurrectionists," as they came to be known, would surreptitiously deliver the corpses to doctors who promised to pay good prices and ask no questions.

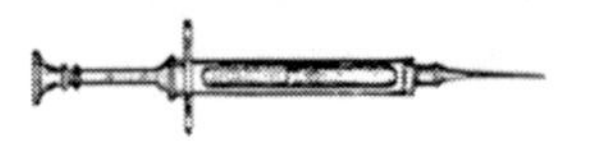

Business was brisk, to the point where outraged local citizens began building walls around cemeteries to protect the graves of their loved ones. Such vigilance made it increasingly difficult for the resurrectionists to practice their trade, but two clever Irish immigrants, William Burke and William Hare, developed a different business strategy. In fact, the new plan actually made life much easier for them, and killed two birds with one stone, as it were. Why go to all the trouble and mess of digging up dirty corpses that had already begun to putrefy, when you could obtain fresh meat for the medical schools by simply creating your own corpses?

It all began innocently enough, relatively speaking. One of Hare's tenants in his boarding house had the nerve to die owing four pounds in rent. Irritated by this loss in income, Hare and Burke decided to conduct a funereal rearrangement for fun and profit. They removed the dead man from his coffin and substituted an equal weight in wood. After the phony funeral, they carried the corpse in a sack to the back door of Dr. Knox's school and were thrilled to receive almost eight pounds!

The lure of easy money then pushed Burke and Hare into more sinister methods of obtaining inventory. One of Hare's tenants fell ill, so why let the poor man linger—and possibly recover, heaven forbid—when they could end his suffering and make another tidy profit? Plying the sick man with whiskey until he was in a stupor, Hare proceeded to pin him to the bed while Burke covered his nose and mouth. Unlike basic strangulation that produced telltale marks on the neck, this method of suffocation left no visible signs of violence, so they were able to deliver unmarked goods to the willing consumers. (This method of homicide later gave rise to the phrase to "burke" a victim.)

The murders continued in a similar pattern—get the victim drunk and suffocate him or her (medical schools needed female corpses, too) and then deliver the bodies in a sack or barrel to Dr. Knox's door. These serial killers practiced their trade with increasing boldness, and only after victim #16, were they apprehended. However, the authorities had little or no evidence with which to try the case, save the one body of #16, so remarkably, they offered William Hare his freedom if he would testify against Burke.

In addition, both men's wives had participated in the grisly crimes in some ways, yet they were not charged and received no punishment. For his testimony, Hare was released. William Burke paid for his crimes at the gallows on January 29, 1829, and fittingly, his body was given to the medical school for dissection. His skeleton remains on display at the school to this day. In another macabre twist, a student stole a piece of his skin and had it made into a wallet! (Payback is a bitch…)

And what of Dr. Knox and the other noble men of medicine who paid for these suspiciously fresh corpses, knowing full well that foul play was somehow involved? With no hard evidence to prosecute, these educated elite of Edinburgh all got off scot-free. (Or should that be Scot-free?)

This is perhaps the most inexcusable part of this gruesome serial murder case—doctors who had sworn an oath to do no harm or injustice, who had devoted their lives to healing the sick, had knowingly and eagerly accepted the bodies of innocent people who had been murdered for profit. Had they not paid Burke and Hare, there wouldn't have been any murders. They were as responsible for these peoples' deaths as if they had burked them all themselves.

Bad Blood

As modern surgery progresses, less invasive techniques are constantly being developed to minimize damage to surrounding tissue and limit the loss of blood. However, things were not always this way. In fact, for about 3,000 years it was quite the opposite—losing blood was considered to be a way to *treat* many ailments.

Bloodletting was already in common practice by the 5th century BC, but probably began centuries earlier in Egypt. In addition to being ancient, it also appears to have been widespread in cultures across the globe. In its earliest forms, it's believed that blood was spilled to release evil spirits out of the body—which was actually probably a lot safer and less dangerous than the practice of trephination, or cutting a hole in the skull to release god-knows-what.

Eventually, the concept of evil spirit-based disease was superseded by a more "scientific" approach, and I do use that term very loosely. As was mentioned earlier in this chapter, in their favor,

  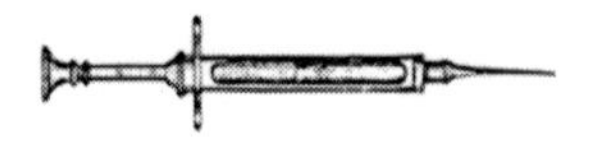

the ancient Greeks began observing patients for biological causes and effects, and felt that opposing forces were at work in the body. Hippocrates concluded that these forces were fluids, or humours, and if they were out of balance then a person would become ill.

The four humours were said to be blood, phlegm, yellow bile, and black bile, and countless nasty practices for "purging" excess humours were developed over the centuries. Sparing the reader the details of the techniques involving both ends of the gastrointestinal tract, we shall focus on relieving the "plethora" of blood that apparently accompanied a variety of maladies.

Does your daughter suffer from epilepsy? Then cut open a vein and drain a pint or two of blood from her little body. Does your wife have female troubles? Then slice a vein in her ankle and she won't lose so much blood, except from her ankle, of course. How about your cousin's various aches and pains? Well, when the proper signs of the zodiac are in the sky, the appropriate veins will be violated to remove all that offending blood from his system. And what about that infection spreading through your foot? Obviously, the best way to relieve the redness and swelling is to drain out so much blood your skin turns white. (Of course, after draining away most of your bacteria-fighting white blood cells, the infection will most likely run rampant and prove fatal. But at least you will present yourself at your funeral as a suitably pale corpse.)

So who, then, in times past was qualified to do all this slicing and dicing? As the profession of physician in the Greek and Roman world disappeared into the dark ages, monks and priests took over the bloody procedures. However, in 1163 AD, Pope Alexander III decreed that the church would no longer allow its representatives to drain their parishioners' blood. (Apparently, members of the clergy were to concentrate solely on draining their parishioners' purses.)

That left the field wide open for a group of men who wielded sharp instruments with great skill, at least with hair—the local barber. Was it such a stretch then, that the same man who inadvertently drew blood when he gave you a shave, would now get paid to cut you in earnest?

The trade of the barber-surgeon flourished for hundreds of years (the official organization in England didn't dissolve until 1744!). One of the most familiar sites in downtown USA traces its origins to this dubious profession. The red and white barber pole came into existence to let the townsfolk know that in addition to a haircut and a shave, the proprietor of that particular establishment would be happy to use his razor to slice open your vein and drain as much blood as you desired. The red color obviously signified the blood, the white represented the bandage or tourniquet, and the pole itself symbolized the stick the patient would clench in his fist to pump up the vessels in his arm—much like making a fist today when giving blood.

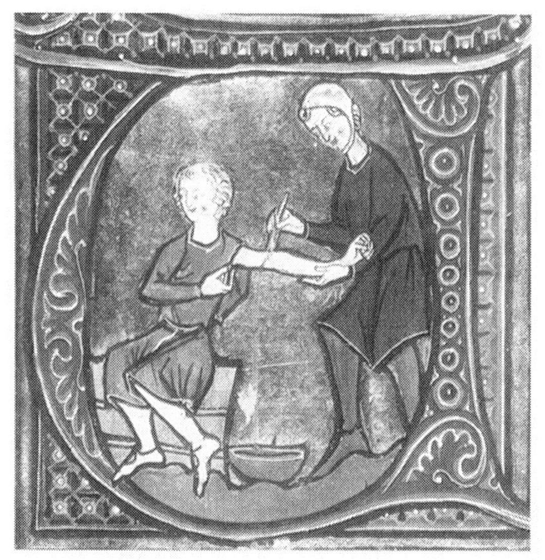

Everyone is smiling in this medieval image of bloodletting.

Some barber-surgeons even went further out on a limb, so to speak, and offered to perform amputations. One can only imagine the results of such an operation being conducted on a patient with no anesthesia, in highly unsanitary conditions, with a razor that had just been used to lance an infected boil on another patient. It might have been more humane to slit the patient's throat...

The tools of the trade evolved over the centuries, from thorns and sharp sticks, to simple knives, to specially designed lancets, to spring-loaded devices that prevented the shaky practitioner from cutting too deep. Various bowls and cups were also designed to catch the vital fluid. Typically, at the point where the patient began to feel faint the procedure would end, which usually translated into relieving the body of one to four pints of blood. Of course, multiple bloodlettings were required to further weaken the victim—I mean

 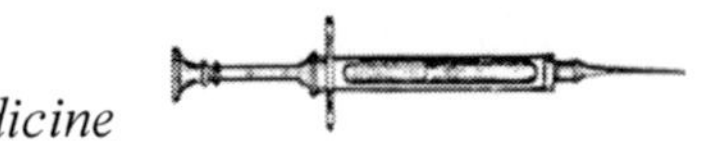

cure the patient—if an adequate response was not obtained the first time.

Perhaps the most infamous example of this deadly repeat performance is the case of America's most beloved figure, George Washington. After coming down with a throat infection in 1799, his doctors decided that the best thing for him was to remove a few pints of blood. Then a few more. And a few more.

It is believed that in a 24-hour period, these quacks extracted a whopping nine pints of blood from our first president. Considering an average adult male has only about fourteen to eighteen pints, it is no wonder that Washington died after most likely losing at least half of his blood. How ironic is it, that after surviving the French and Indian War, the Revolutionary War, and the most dangerous foe of them all, the U.S. Congress, he was finally done in by his doctors!

It's generally agreed that by the late nineteenth century, the practice of bloodletting finally ended…or did it? Remarkably, a CNN report in 1998 revealed that people in India are still undergoing the procedure. They believe that losing their bad blood will cure everything from arthritis to cancer. Of course, as we have seen, if you drain enough blood it will cure everything—permanently!

Unfortunately, the depth of human ignorance appears to equal the vast oceans of blood that have been spilled in vein (sorry, couldn't help myself) across the centuries. How many lives were cut short because of bloodletting? It is astounding that it took thousands of years for this practice to be recognized for what it was—an expression of humour that was no laughing matter…

If someone was very pale, they were fed red foods, or given pink or red medicine. In that same vein, so to speak, patients were given blood to drink, which seemed to work wonders and prove the belief that colors could cure. Of course, blood relieved the symptoms of anemia because of its iron content, but why spoil a good myth with science?

"If excessive smoking actually plays a role in the production of lung cancer, it seems to be a minor one."

W.C. Heuper, National Cancer Institute, 1954

Lying About Your Age

People have probably been lying about their ages since civilization began. Men and women have claimed to be younger to appear more attractive to the opposite sex. Teens have claimed to be older to join the army, drink, and appear more attractive to the opposite sex. It is not an act of Bad Science to lie about your age—unless you happen to be part of a medical study.

In the Andes of Ecuador in 1970, scientists began studying the inhabitants of the small town of Vilcabamba. Many of the people of the town were supposedly quite elderly, but still vigorous and healthy, and the researchers wanted to see if there was any link between their diets, lower heart disease rates, and their remarkable longevity.

However, they were unprepared for just how remarkable the claims of longevity were—of its 819 residents, 9 asserted that they were over 100 years old! That percentage of centenarians was 366 times greater than in the U.S. at the time. These people also weren't just 100, or 101, or even 105. One man claimed to be 123, another 132, and several to be over 140, and they all still worked on their farms every day! Further research then discovered an additional 14 centenarians. These seemed to be impossible numbers, but a search of the town's birth and baptismal records appeared to substantiate these claims.

The "Shangri-La of the Andes" suddenly became a magnet for tourists and more researchers trying to discover the Fountain of Youth from which so many residents obviously drank. Articles and books were written about Vilcabamba's "longevos" (very old people), and studies were conducted that concluded it was the charged mountain air, the constant exercise, the food, or all of the above that kept the people alive and well.

While there was no doubt that there was a high percentage of elderly people in Vilcabamba, and that they were quite active in their golden years, all of the initial research failed to take into account one vital factor—the people were lying about their ages! It seems that in this town, it was something of a point of pride to reach old age, so people being people, they simply exaggerated how old they were. For example: in 1971, a man said he was 122, but just

three years later, he reported that he was 134. Then there was the case of Miguel Carpio Mendieta who was 61 in 1944, but claimed to be 70. Five years later he said he was 80. In 1970, he was actually 87, but by then reported his age to be 121, and three years later declared himself to be 127!

Vilcabamba, which translates as "Town Where Old People Lie Through What Few Teeth They Have Left."

In addition to the prevailing culture of age exaggeration, scientists had been fooled by the birth and baptismal records, because so many people had the same names. Researchers had mistaken people's fathers, grandfathers, and other relatives' records as those of their much younger descendants. Also, the reason that there was such a high percentage of elderly people in town was simply that the young people had moved away!

After a more careful examination of the records and residents, it turned out that *NO ONE* in Vilcabamba was over 100 years of age! Further, in reality, the life expectancy was actually *lower* than in the U.S.!

Despite all the evidence to the contrary, today Vilcabamba still proclaims itself to be in the "Valley of Longevity" where "years are added to your life…and life is added to your years!" Tourists still visit hoping to find the secret to long life, and they drink large quantities of the town's spring water—just in case there really is some truth to it being a Fountain of Youth.

Of course, people could just try to live healthy lifestyles regardless of where they reside, but why bother to go to all the

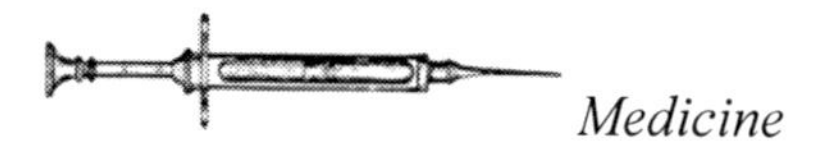

trouble of personal responsibility for one's health, when you can look for a quick fix by traveling to a remote town in the Andes?

The Pen is Mightier than the Scalpel

While great strides in surgical knowledge have allowed organ transplants to become commonplace, the battle for survival does not end with the operation. Organ rejection is the ever-present danger for the recipient, and the search still continues for a course of treatment to coax the body into acceptance, without significantly compromising its natural defenses.

In 1973, Dr. William Summerlin of the prestigious Sloan-Kettering Institute for Cancer Research, claimed to have invented the solution; literally, a solution which eliminated rejection. For his alleged discovery, Dr. Summerlin seemed destined to add his name to the annals of medical history. In actuality he was, but not quite in the manner he had hoped. There was one teeny-weeny little problem: Dr. Summerlin faked his experiments.

The experiments he claimed to have conducted with his new solution involved transplanting pieces of skin from black mice to white mice. The white mice did indeed begin exhibiting black patches of skin with no signs of rejection. The results appeared remarkable.

Of course, appearances can be deceiving. Many laboratories attempted to reproduce Dr. Summerlin's results, but every one failed. The problem with these other labs, as it turned out, was not a lack of the right medical techniques, but simply a lack of the right office supplies. Dr. Summerlin had cleverly chosen to claim to have transplanted the black skin to the white mice because his "surgery" did not involve a scalpel. His instrument of choice was a black felt-tip pen.

The fraud was exposed one day when an assistant discovered the good doctor carefully coloring in patches of the white mice's skin with the pen. For this outrageous act of deceit, Summerlin was suspended, but not fired (although he never returned to the Institute). And despite lying to colleagues, fabricating results, and coloring

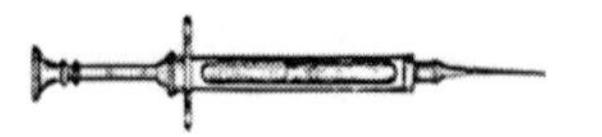

rodents without a license, he was actually allowed to continue to practice medicine!

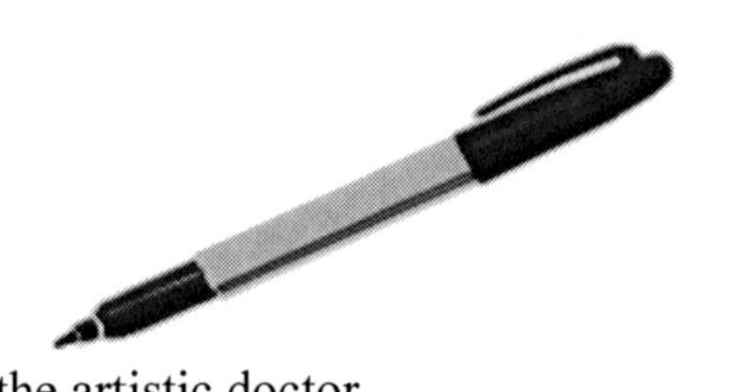

One cannot help wondering if his future patients became concerned as moles and freckles began appearing after their visits to the artistic doctor.

When Good Scientists Go Bad

It's bad enough when people with questionable credentials perpetrate fraud or a hoax, but it is arguably even worse when good scientists go bad. But why should someone who is smart, educated, and talented travel down the road of Bad Science?

Unless you have worked in a lab in a university, research facility, or in industry, you can't fully appreciate the pressure that is applied to produce results. There is the pressure to publish papers, make new discoveries, make products that make money, and the need to constantly justify your position. All this pressure is too much for some people, and they succumb to the temptation to sidestep the rigors of true science and embrace the dark side of bogus results.

As a prime example, Dr. John Darsee was once lauded as "clearly one of the most remarkable young men in American medicine. It is not extravagant to say that he became a legendary figure during his year as chief resident in medicine at Grady Memorial Hospital." This ringing endorsement came from a cardiologist at Emory University, and it helped Darsee to land a coveted research fellowship at Harvard Medical School in 1979.

Darsee hit the ground running with his research on the effects of heart medications on dogs, and in just two years he already had seven of his papers published in major journals. He was brilliant, hard-working—often putting in 90-hour weeks—and he was on the fast track to a position on the faculty…until someone noticed something peculiar.

Just a few minutes after the start of an experiment, three colleagues noticed that Darsee was filling in data for an entry labeled "24 seconds."

That was as it should be, but then Darsee filled in the data for "72 hours."

Wait a minute, how could he do that?

Moments later, he moved on to the entries for "one week" and "two weeks," even though the experiment was only a few minutes old.

In other words, Darsee was creating science fiction, not recording science fact. When confronted, he admitted the pressure had gotten to him so he made up all the results of this one study, but he swore this was the *only* time he had *ever* done such a thing.

In reality, this was the *only* time he had actually been caught fabricating data, and the whole mess quickly hit the fan. His supervisors and other researchers spent months of precious time uncovering other cases of Darsee's creative data, time that could have been spent trying to save lives. A hospital had to return over $122,000 that the National Institute of Health had given them for one of Darsee's studies, which turned out to be fraudulent. Careers were forever tainted as a shadow of doubt fell upon anyone who had ever worked with Darsee. One man's weakness led to trouble for scores of innocent, dedicated professionals.

Needless to say, the faculty position was taken off the table, his fellowship was terminated, and Darsee had to leave the field of research. He became an instructor, where hopefully, he didn't assign grades for all of the quizzes, the midterm, and final exams during the first class.

A Hole in One

When examining mankind's violent history, the human skull surfaces as the Rodney Dangerfield of body parts; it has gotten no respect. From primitive bone clubs, to medieval maces, to Louisville Sluggers, men have been bashing one another's skulls to bits since time immemorial. Our resourceful ancestors also employed their skills to perfect swords and axes for the sole purpose of removing

the entire skull clean off the shoulders. And who can forget the guillotine, an ingenious time-saving device which deprived more bodies of their skulls than you could shake a loaf of French bread at.

However, not all cranial violations were meant to cause harm. An ancient practice known as trephination actually involved gouging a hole through the skull for the patient's own good. Whether intended to expel evil spirits or simply relieve pressure on the brain, trephination was performed around the world for thousands of years. Yet despite the good intentions of the healers, the procedure was gruesome: after slicing open the scalp to expose the skull, a crude saw slowly dug a series of grooves until a chunk of skull could be pried out. More remarkable than the patients' abilities to withstand these grisly operations, is the fact that many survived for years afterward (as is evidenced by the new growth of bone around the holes).

Skulls are still being cut open today, but fortunately the procedures are being carried out under anesthesia by skilled brain surgeons in the sterile environments of operating rooms. But then again, not always.

In 1962, Dr. Bart Hughes of the Netherlands believed that LSD was a wonderful drug, because it increased the volume of blood to the brain. In addition, this "Brainbloodvolume," as he cleverly termed it, could also be increased by liberating the brain from its bony prison. In an attempt to prove his theory, Hughes drilled a hole in his own head. Apparently, the self-inflicted head wound didn't do much to hone his intellectual skills, because when he announced his discovery to his fellow countrymen, he fully expected to be hailed as a hero. In fact, they hailed him as a lunatic and put him in an asylum.

A few years after he got out, Hughes actually managed to get a disciple, an Englishman, Joseph Mellen. Mellen later wrote a book about his experiences with Hughes and trephination, aptly named *Bore Hole* (although it probably would have been more appropriate to name it *The English Mental Patient,* or *Mellen's Melon*). The book begins with the line, "This is the story of how I came to drill a hole in my skull to get permanently high," which goes a long way to explain the rest of the book.

Mellen describes his early years and how he became an accountant. After experimenting with drugs, including LSD, he met

Hughes and was faced with the dilemma of continuing to be an accountant, or drilling a hole in his head. As the latter seemed less painful in the long run, he bought a circular skull drill and prepared to release his brain into higher consciousness. Taking a tab of LSD to bolster his resolve (and suppress any shred of common sense), Mellen sliced open his scalp, but was greatly disappointed when he found that he was unable to jam the point of the drill bit into his own head. Since British officials decided to keep Dr. Hughes on his side of the North Sea, Mellen had to wait until his girlfriend, Amanda, returned from a trip to assist him in his great liberation.

By throwing all her weight against the drill bit, Amanda finally got it to dig into her boyfriend's skull. Then she dutifully sat by and watched until he drilled himself into unconsciousness. After recovering in the hospital and spending time in jail for possession of drugs, the undaunted Mellen tried a third time to drill a hole in his head. And as the saying goes, the third time is the charm, as his own description of the event illustrates:

After some time there was an ominous sounding schlurp and the sound of bubbling. I drew the trepan (bone drill) *out and the gurgling continued. It sounded like air bubbles running under the skull as they were pressed out. I looked at the trepan and there was a bit of bone in it. At last!*

However, the piece of bone was a small one and Mellen later decided it wasn't big enough. All alone this time, he pressed an electric drill against his skull, but the bit wouldn't penetrate and the drill burned out after half an hour. Quickly getting the drill repaired, he tried again the next day:

This time I was not in any doubt. The drill head went at least an inch deep through the hole. A great gush of blood followed my withdrawal of the drill. In the mirror I could see the blood in the hole rising and falling with the pulsation of the brain.

According to Mellen, the self-trephination was a rousing success. It was so successful, in fact, that Amanda decided to drill a hole in her own head, while filming the entire procedure. (During screenings of her film, *Heartbeat in the Brain,* people often fainted, "dropping off their seats one by one like ripe

 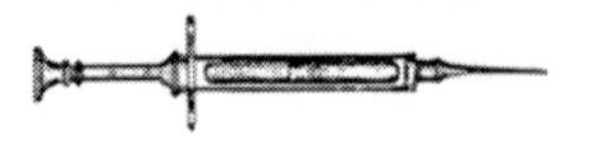 

plums.") Not only did Amanda and Mellen both survive, they had a child and went on to have successful careers in the art field (specializing, no doubt, in the works of Van Gogh and other self-mutilators).

In the past few decades, there have been isolated reports of others undergoing trephination, but fortunately, mere body piercing seems to have become the self-mutilation procedure of choice of today's younger generation. But who can tell what the future will bring? As medical costs continue to soar, what enterprising health insurance executive wouldn't dream of inexpensive home brain surgery kits? And even with limited Brainbloodvolume, would it be too hard to imagine a 22nd-century Bob Vila, popping a fresh trepan onto his drill each week and featuring a different brain refurbishing technique on his show *This Old Skull*?

Or, should we instead hope for the day when medicine conquers all illnesses, and mankind finally evolves to a point where we conquer the urge to bash one another's skulls to bits?

Now *that* is far-fetched.

Human Guinea Pigs

Human experimentation is not just the stuff of Nazi concentration camps and conspiracy theorists. While it's bad enough to be forced against your will to undergo medical experiments, it is arguably worse to be subjected to potentially harmful substances and procedures without your knowledge.

Sounds like fiction, or the work of an evil foreign dictator? Unfortunately, no. We need look no further than Uncle Sam, American doctors, and our own backyards. While the mentally ill, the sick, children, minorities, and prisoners were the most frequent targets of non-consensual experimentation, the general public was in the medical testing crosshairs, as well. Presented for your disapproval, a brief history of human guinea pigs in the United States of America:

1913: In Pennsylvania, 146 children were inoculated with syphilis and 15 children had their eyes tested with tuberculin, causing permanent blindness in some of them.

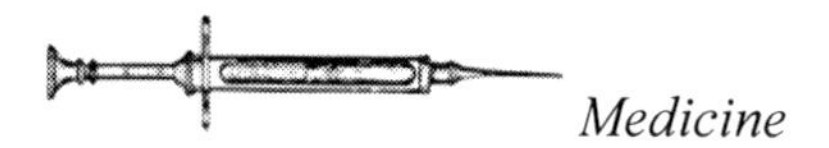

1927: A woman in Charlottesville, Virginia was legally sterilized against her will. The reason given: she was "mentally normal," but her mother was "mentally retarded" so there was the potential that she could have a "less than normal child."

Laws allowing forced sterilizations of the mentally ill and their normal daughters spread to 17 states, and the grounds for such procedures grew to include women who were criminals, prostitutes, and alcoholics. The law wasn't repealed in Virginia until 1981.

In the 1960s and 70s, there was a similar campaign of non-consensual sterilizations of Native American women, resulted in tens of thousands of women—some as young as 15—being unknowingly or forcibly sterilized.

1931: Dr. Cornelius Rhoads, while working under the auspices of the Rockefeller Institute, injects cancer cells into Puerto Ricans, causing several deaths. Rather than facing prosecution, he runs chemical warfare projects and establishes several biological warfare facilities for the U.S. Army. As a member of the U.S. Atomic Commission, he conducts radiation experiments on soldiers and civilians.

Dr. Rhoads was awarded the U.S. Legion of Merit for his work.

1932: The infamous Tuskegee Syphilis Study begins. From 1932 to 1972, 399 poor African-American men with syphilis are studied for the effects of the disease. Even after the discovery of penicillin, treatments are withheld. Not only do all these men suffer the ravages of the disease, but they spread it to women and children.

1942: The army begins experiments with poison gas on thousands of soldiers.

1945: The army begins radiation experiments on soldiers. In the years following, other experiments include feeding radioactive substances to mentally impaired children.

 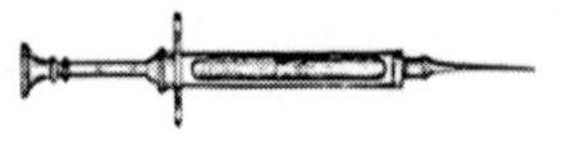

1947: The CIA begins experimenting with LSD and other mind-altering drugs on soldiers and civilians, without their knowledge.

1949: The CIA, Department of Defense, and various branches of the military conduct years of open air biological and chemical warfare testing on unknowing civilians across the country.

In 1977, Senate hearings confirm that between 1949-69, at least 239 populated areas had been exposed to biological agents, including cities such as San Francisco, Washington, D.C., Key West, Panama City, Minneapolis, and St. Louis.

The extent of the effects of these experiments may never be known.

This is just a very brief list of what has become public knowledge, and one shudders to think what secrets still exist. Has all experimentation by the U.S. government on its citizens and military personnel ceased, once and for all?

The answer depends upon how gullible and naïve you are…

Fun with Radium and X-Rays

As various forms of radiation were known to be composed of high-energy particles, and people often suffered from fatigue, it would make sense that radiation could give people more energy, right?

Unfortunately, this was a commonly held belief that led to some dangerous "cures." For example, why not keep a special dispenser of "Radium Water" on the kitchen table so junior can revitalize himself after a long day at school? Radiated water was very popular in the early 20th century and was at its peak in the 1920s. Some radioactive jugs imparted their alleged health benefits to the water placed in them, while other "remedies" came pre-irradiated, such as *Radithor*, which was manufactured in New Jersey.

Radithor was billed as "A Cure for the Living Dead," referring to the mentally ill who were trapped inside their own minds, but it was also purchased by the general health-conscious public. For example, a Pittsburgh steel tycoon, Eben Beyers, consumed 1400 bottles. However, Beyers stopped drinking *Radithor* after the subsequent radium poisoning required the surgical removal of parts of his mouth and jaw, which was shortly followed by his death.

Then there were radium creams, toothpastes, radium bags placed over rheumatic joints, uranium blankets for arthritis, and a digestive remedy containing thorium. However, the *pièce de résistance* was unquestionably the Vita Radium Suppositories (otherwise known as "the little pills that might as well make you bend over and kiss your ass goodbye.")

The Vita Radium brochure proclaimed:

Weak Discouraged Men!
Now Bubble Over with Joyous Vitality
Through the Use of
Glands and Radium

"...properly functioning glands make themselves known in a quick, brisk step, mental alertness and the ability to live and love in the fullest sense of the word...A man must be in a bad way indeed to sit back and be satisfied without the pleasures that are his birthright! Try them and see what good results you get!"

A man must also be in a bad way to stick a radioactive suppository up his butt in an attempt to have sex!

People didn't just limit themselves to having fun introducing radioactive substances into their bodies; they got to play with x-rays, too. Invented in the early 1900s, shoe-fitting fluoroscopes helped mothers determine if their children's new shoes were fitting properly. In the interest of healthy feet, deadly x-rays were dished out in abundance.

The most popular of these shoe store gimmicks was manufactured by the Adrian X-Ray Company of Milwaukee, WI, and was designed by Brooks Stevens, who also gave the world the Wienermobile. In England, the Ped-O-Scope was the public x-ray machine of choice. By the early 1950s in the United States alone, over 10,000 shoe-fitting x-ray devices across the country were zapping customers and sales people. It will never be known how many cancers these devices provoked, but there was at least one report of a shoe model losing her leg due to radiation burns.

[Caption should read: Hey, let's all gather around the Ped-O-Scope and watch little Sally get cancer!]

As the hazards of poorly shielded x-ray machines in the middle of stores became better known, increasingly stricter regulations were imposed, until most machines were shut down by 1970. However, as late as 1981, one device was still in operation in a department store in West Virginia!

As in many cases of bad medicine, people desperate for relief from what ailed them led to many falling prey to the ignorant and greedy, who indiscriminately peddled radioactive remedies to men, women, and children. We may marvel at their stupidity, but just turn on any infomercial late at night to realize that snake oil is still snake oil, just with fancier packaging and slicker marketing techniques.

Such people should be punished with Suppositories of Truth. I have no idea what those would be, but I bet they would sting going in and make one repent coming out.

VITA RADIUM SUPPOSITORIES

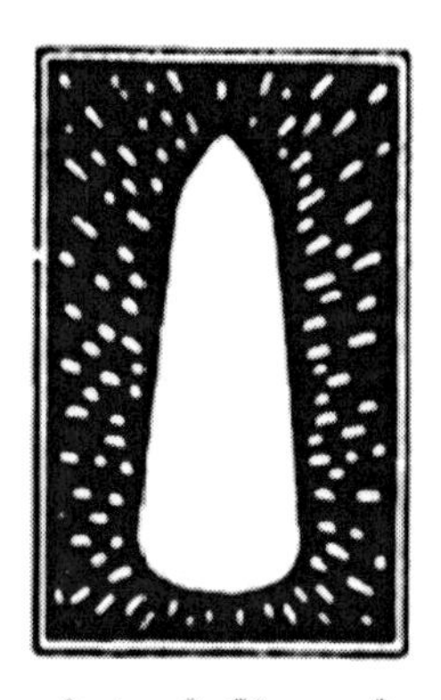

Actual Size of Suppository

OUR VITA RADIUM SUPPOSITORIES (HIGH STRENGTH) are one of the outstanding triumphs of Radium Science. These Suppositories are guaranteed to contain REAL RADIUM—in the exact amount for most beneficial effect. They are inserted per rectum, one each night, this being one of the several practical and successful ways of introducing Radium into the system.

After insertion, the Suppository quickly dissolves and the Radium is absorbed by the walls of the colon; then, within a few minutes, it enters the blood stream and traverses the entire body. Every tissue, every organ of the body is bombarded by its health-giving electric atoms. Thus the use of these Suppositories has an effect on the human body like recharging has on an electric battery.

And remember, Radium taken into the system remains for months, continuing its curative, restorative work. Thus, the effects are NOT merely temporary.

VITA RADIUM SUPPOSITORIES are guaranteed to be non-injurious—they are perfectly safe for anyone to use. Their action is due solely to the Radium contained therein.

"X-rays will prove to be a hoax."

Lord Kelvin, mathematician and physicist, president of the British Royal Society 1895

"Heavier-than-air flying machines are impossible."

Lord Kelvin, 1895

"Radio has no future."

Lord Kelvin, 1897

 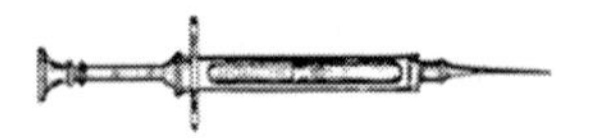

According to repeated nationwide surveys,

More Doctors Smoke CAMELS than any other cigarette!

Doctors in every branch of medicine were asked, "What cigarette do you smoke?" The brand named most was Camel!

You'll enjoy Camels for the same reasons so many doctors enjoy them. Camels have such cool mildness, pack after pack, and a flavor unmatched by any other cigarette.

Make this sensible test: Smoke only Camels for 30 days and see how well Camels please your taste, how well they suit your throat as your steady smoke. You'll see how enjoyable a cigarette can be!

THE DOCTORS' CHOICE IS AMERICA'S CHOICE!

[illegible]

[illegible]

[illegible]

Chemistry and Pharmaceuticals

The Sears Catalog *Did* Have Everything

For generations, Americans purchased everything from overalls to corsets, and appliances to complete homebuilding kits from the Sears catalog. For families in rural locations, mail order was essential to having access to goods they might otherwise never have the opportunity to buy. Of course, they weren't all practical items. What child growing up in the 1960s and 70s could forget the day the tantalizing Sears Christmas Wish Book arrived in the mail—filled with every toy your little heart could desire.

Yes, it seemed that at one time or another, you could find anything in a Sears catalog…like morphine…and heroin…

In 1894, a Bayer Company chemist, Heinrich Dreser, developed diacetylmorphine. The claim was that it was a "heroic" drug that would work as effectively as morphine, but would free people from addiction. Based upon this noble, but oh so erroneous, premise, Bayer christened the drug Heroin and began widespread distribution. Unfortunately, in reality, heroin is several times more addictive than morphine, but who was counting?

Actually, someone was counting—the resourceful salesmen at Sears who offered heroin in their catalogs. In fact, they had a special drug catalog in 1902 that not only offered heroin, but morphine, opium, and marijuana, as well. That was quite a Wish Book for the estimated one million opiate addicts in the United States at that time.

In an ironic twist, some Sears catalogs even offered an infallible cure to break morphine and opium addiction, for a mere 75 cents. And if that didn't work, you could always buy more drugs for just a dollar.

Arguably their best marketing strategy, and possibly an instant bestseller, was the morphine concoction designed for lonely housewives—not for them to take themselves, but for these unfortunate wives to surreptitiously add to their husbands' drinks at dinner so as to render them in such a diminished state of consciousness, that they would be incapable of going out that night and getting into some kind of mischief.

Alas, such drugs as those for keeping your husband at home are no longer available through mail order, but isn't that what Victoria's Secret catalogs are for now?

> "The more important fundamental laws and facts of physical science have all been discovered, and these are now so firmly established that the possibility of their ever being supplanted in consequence of new discoveries is exceedingly remote."
>
> Albert. A. Michelson, physicist, 1894

Legal Poison

Life before the discovery of antibiotics such as penicillin was difficult, as even minor infections could spread and become deadly.

When sulfa drugs were developed and found to be useful in disrupting the reproduction of bacteria, they were hailed as something of a miracle. Available in tablet and powder form, they clearly saved lives that otherwise would have been lost.

In 1937, salesmen at the Massengill Company suggested developing a liquid form of sulfanilamide, as customers in their territories had requested something that tasted good, and therefore might be easier to administer to patients, such as young children, who were reluctant to swallow pills. The task fell upon the company's chief chemist, Harold Watkins. What Watkins needed to do was find a way to dissolve the drug in a raspberry-flavored syrup. Trial and error finally revealed that sulfanilamide dissolved quite nicely in diethylene glycol, and created a liquid that looked, smelled, and most importantly, tasted good.

Bottles of Elixir of Sulfanilamide

Unfortunately, there was still one glaring error—safety. No one at Massengill thought to test the new Elixir of Sulfanilamide to see if it was actually safe to ingest! The substance Watkins had chosen as a solvent—diethylene glycol—is now better known as the main ingredient in your car's anti-freeze, and it is nothing short of a poison. Although at the time, there were published reports indicating diethylene glycol's toxic effects on the human body, e.g., causing kidney failure, no one at Massengill bothered to look them up.

No less than 240 gallons of the deadly Elixir of Sulfanilamide was distributed across the country in 633 separate shipments in early September of 1937. People were delighted by the sweet raspberry taste. Then they started dying.

 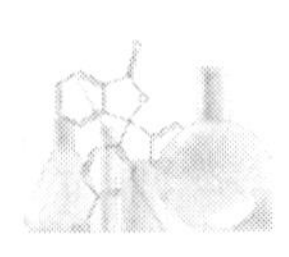

On October 11, physicians in Tulsa, Oklahoma, reported to the American Medical Association that nine patients (eight of whom were children) who had been prescribed the Elixir had died. Massengill was contacted and the diethylene glycol was discovered to be the culprit. Not until the 14th of October did a physician inform the Food and Drug Administration, which remarkably under the law, was powerless to prohibit the sale of toxic substances.

Massengill officials informed the FDA that the situation was under control, as they had sent out over a thousand telegrams asking that salesmen, doctors, and pharmacists please return the Elixir. However, they didn't bother to mention that the medicine was poisonous! The FDA then twisted a few arms, and Massengill finally sent out telegrams with a more urgent message: "Imperative you take up immediately all elixir sulfanilamide dispensed. Product may be dangerous to life. Return all stocks, our expense."

As the Elixir had already spread like an unchecked infection across the country, into many rural locations, this task was easier said than done. Traveling salesmen were hard to track down, drugstores did not keep detailed records of prescriptions, and doctors actually lied for fear of prosecution. One doctor claimed he had prescribed the drug to five people, but all of them were fine. Upon further investigation, however, the FDA inspector discovered that, in reality, all five people had died!

One distraught physician, a Dr. Calhoun, who had been practicing medicine for over 25 years, lamented that six of his patients had died, one of whom was his best friend. In a poignant letter, he wrote that the realization that the medicine he trusted and had prescribed had killed these innocent people, "has given me such days and nights of mental and spiritual agony as I did not believe a human being could undergo and survive. I have known hours when death for me would be a welcome relief from this agony."

Slowly but surely, the lethal drug was rounded up, but not before causing a staggering loss of life. At least 358 people were found to be poisoned by the tasty raspberry concoction, and 107 of them died. Additionally, one more death can also be attributed to this tragic case—chemist Harold Watkins committed suicide when he learned of his terrible mistake.

Too bad the company's owner, Dr. Samuel Massengill, didn't feel similar remorse. In a statement worthy of any modern-day corporate official trying to cover his company's ass, Massengill declared, "My chemists and I deeply regret the fatal results, but there was no error in the manufacture of the product. We have been supplying a legitimate professional demand and not once could have foreseen the unlooked-for results. I do not feel that there was any responsibility on our part."

"No error!?"

"Not once could have foreseen the unlooked-for results!?"

"I do not feel that there was any responsibility on our part!?"

Unbelievable, but the story could have been much worse. As the FDA had no legal right to seize the product just because it happened to be poisonous, they only managed to save the public from further deaths by recalling the Elixir of Sulfanilamide because it was mislabeled! As the term "elixir' is defined as a solution of alcohol and this product used diethylene glycol, the FDA was able to act upon this technicality.

At least these unfortunate people did not die in vain—the uproar over the case finally led to the more stringent Federal Food, Drug, and Cosmetic Act of 1938, which finally gave the FDA the power to protect the public from such lethal products. Of course, it can be effectively argued today that many food additives and pharmaceuticals are still toxic, and politics and money are often cited as the means for companies to retain profits at the expense of the public health.

This case and the current state of affairs bring to mind two old sayings:

"Buyer beware," and "Pick your poison."

It is also brings to mind that Massengill is best known today for their disposable douches.

Defective Testing

In 1954, the German company, Grünenthal, secured a patent on a new "wonder drug," which acted as a sedative and was found to be

very effective in reducing the awful symptoms of morning sickness. Tens of thousands of pregnant women around the world started taking the drug. While it was marketed under many different brands, the name was commonly known as thalidomide—a name that will now live in infamy as one of the worst medical tragedies in history.

Even after doctors started reporting an apparent link between the use of thalidomide and birth defects, greedy manufacturers stepped up their advertising campaigns for this "harmless" drug. Finally in 1961, countries began banning thalidomide, but it would be another full year before the entire world realized that this drug was the cause of the babies being born with seriously shortened, malformed limbs.

Unfortunately, the damage was already done, as an estimated 10,000 to 20,000 children worldwide had been born with major thalidomide-related defects and problems. In addition to the limb deformities, the drug also caused abnormalities in the eyes, ears, lips, mouth, heart, genitals, kidneys, digestive tract, and nervous system.

Lawsuits were filed against Grünenthal, but in their defense they claimed the drug had been thoroughly tested on "approximately 10 strains of rats, 15 strains of mice, 11 breeds of rabbits, 2 breeds of dogs, 3 strains of hamsters, 8 species of primates, and in other such varied species as cats, armadillos, guinea pigs, swine and ferrets," and adverse effects had "been induced only occasionally."

Even though it had been proven that animal testing *can never fully predict reactions in humans*, the courts found in favor of Grünenthal, and they got off scot-free. Even more remarkable, thalidomide is still in use, although for conditions other than morning sickness.

Today, when people complain that their prescriptions are too expensive, pharmaceutical companies counter by complaining that it's because they are required to put potential new drugs through a long and expensive series of tests and trials. Perhaps everything does cost too much, but no mere financial concerns can ever top the cost of medical mistakes such as thalidomide.

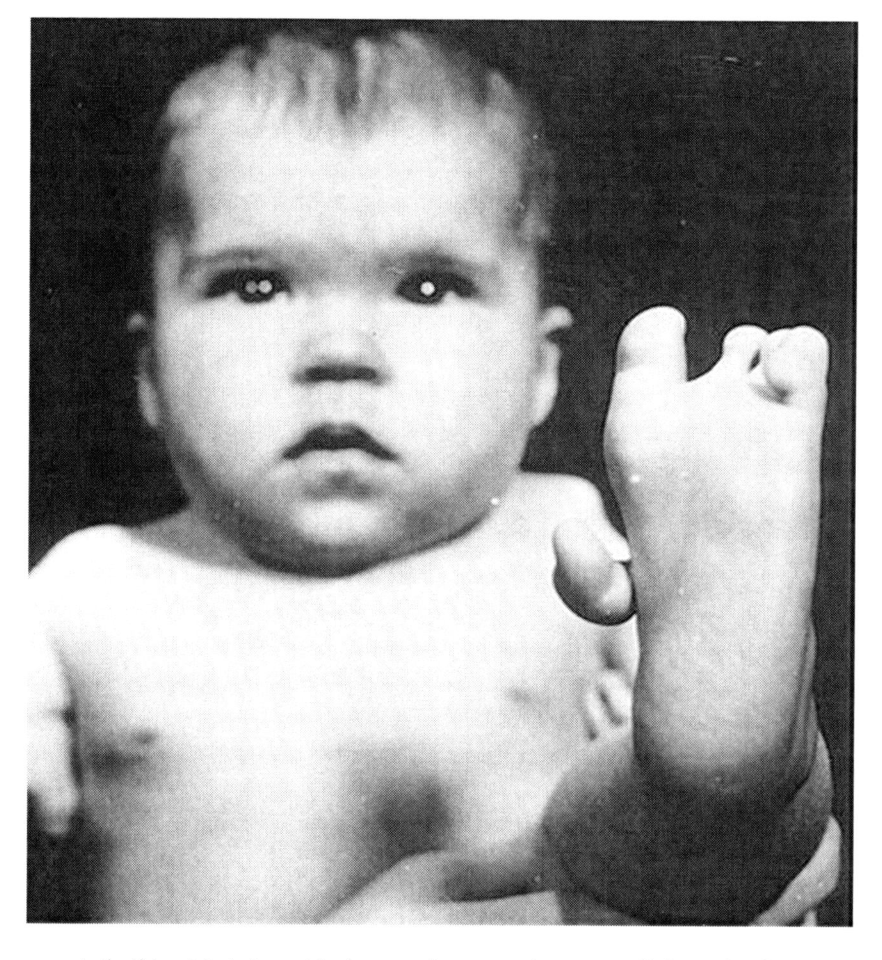

A thalidomide baby with shortened arms and an extra digit on the foot.

Nobel Prize for Death and Destruction

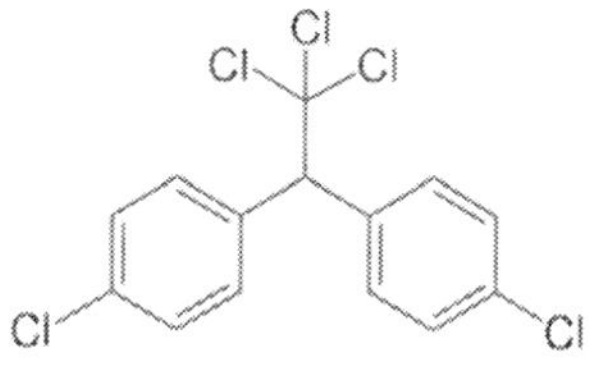

Dichlorodiphenyltrichloroethane was first synthesized in 1874, but it would be sixty-five years before it was discovered that it could be used as an effective pesticide. The chemical was used extensively during World War II to combat mosquitoes and other disease-carrying insects. In 1948, chemist Paul Hermann Müller was

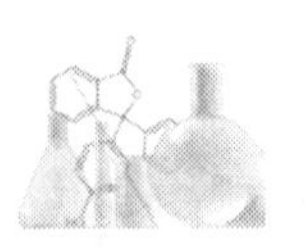

awarded the Nobel Prize in Physiology or Medicine "for his discovery of the high efficiency" of this chemical "as a contact poison against several arthropods."

This miracle chemical, which was to better the lives of mankind, was also known as DDT.

1947 ad in *Time* magazine.

Perhaps the Nobel Prize committee should have studied the long- term effects of introducing massive quantities of DDT into the ecosystem. Perhaps driving down the street spraying DDT directly onto lawns and onto the children playing on them wasn't such a good idea. Perhaps using 1.35 billion pounds of DDT in the United States alone, wasn't the best thing for fish, birds, animals, and humans.

In 1962, Rachel Carson wrote the groundbreaking book, *Silent Spring*, about the dangers of pesticides such as DDT. The book launched official investigations and raised awareness about what we were doing to the environment. Evidence mounted that DDT could be the cause of birth defects and cancers in people, as well as decimating certain animal species such as the bald eagle. Most uses of the pesticide were finally banned in 1972 in the United States, although it is still used in 11 countries, and some scientists still argue for its use in this country.

DDT is being sprayed on children playing on a beach in Long Island in 1945.

The human and environmental toll exacted by DDT can never be fully measured. However, one thing is certain—you can be sure that around the world, equally dangerous chemicals are being introduced into our soil, water, air, and food supply, all in the name of science, progress, and profit.

You Are What You Eat?

It has often been said you are what you eat, which does have some scientific basis. For example, if you eat an unhealthy diet full of fat, sugar, additives, and artificial ingredients (which constitute the four basic food groups according to most Americans), you are likely to be an unhealthy person. However, people often took this saying a bit too literally.

Several ancient cultures thought that by eating the heart of your enemy you would gain his strength and courage. (Of course, if your

 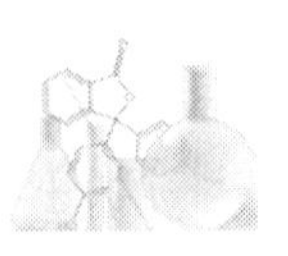

enemy is dead at your feet with his heart ripped out, you are naturally going to feel emboldened.) Unfortunately, a similar practice was not so ancient—during World War II, Japanese soldiers killed and ate five captured American pilots on Chichi-jima (150 miles north of Iwo Jima), and there were many other reports of Japanese cannibalism throughout the war. Livers were particularly sought-after.

When humans weren't eating other humans, they were eating other animal's body parts to try to acquire certain attributes. For example, it was believed that since foxes could run for long distances, eating their lungs would help cure lung disease. Bears were very furry, so eating bear fat could cure baldness. It was also believed that if you munched on a lion's heart it would make you, well, lion-hearted.

Then there are the rhinoceros horns and tiger penises. Unfortunately, even today there is the belief that for those men who are "erectile-challenged," consuming the ground up horn of a rhinoceros will help you get your own horn up. The same goes for eating a bowl of tiger penis soup, which can sell for as much as $350 on the black market in Southeast Asia! Unfortunately, this ridiculous practice has helped drive these poor creatures onto the endangered species list. But hey, what's the extinction of a few magnificent animal species compared to some stupid old Asian guy trying to get an erection?

Fortunately, where extinction awareness campaigns have failed, a little pharmaceutical wonder pill might succeed—sildenafil citrate, a.k.a., the "Little Blue Pill," a.k.a. Viagra®. This magic bullet erectile dysfunction pill just might keep real bullets from taking down more rhinos and tigers, as even at about $30 per pill, it's a lot cheaper than a steaming bowl of tiger penis soup!

Of course, god forbid someone actually takes responsibility for his own health, the safety of other species, and the welfare of the planet, by eating a clean vegetarian diet and getting regular exercise!

Pardon that last line, I know I'm just talking nonsense now. What I should have told the readers in the spirit of "You are what you eat," was by all means, grab a bag of greasy chips and go plant yourself as a couch potato!

One Bad Apple

Perhaps no other scientist in history ever enjoyed the demigod status bestowed upon Sir Isaac Newton. This feeling of reverence is clearly demonstrated in Alexander Pope's rhapsodic lines:

"Nature and Nature's laws lay hid in night:
God said, Let Newton be! and all was light."

While Newton's contribution to science was staggering, there were a few dark, little corners in his life where that light apparently didn't penetrate.

A portrait of Newton from 1702.

Newton is often considered to be one of the first of the modern scientists, yet John Maynard Keynes believed that Newton was actually "the last of the magicians"—associating him with a line of mystical brotherhoods dating back to the days of ancient Babylonia. Keynes took this radically opposite view after studying some of Newton's papers he purchased in an auction in 1936. The papers were a small part of a vast body of works Newton had written on alchemy.

Although many alchemists became the legitimate ancestors of chemistry (albeit, often by accident), much of alchemy was shrouded in secrecy to protect what its adherents believed to be the mystical answers to the riddles of the universe. Some of those riddles involved the planets, but they weren't studied for any astronomical reasons. It was believed that each of the heavenly bodies was connected to one of the seven special metals and that transmutations (like turning lead into gold) were most easily

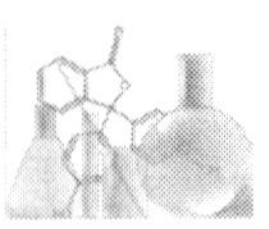

accomplished under the beneficent influence of the appropriate planet. The planets and their metals were as follows:

Sun - Gold
Moon - Silver
Mercury - Quicksilver
Venus - Copper
Mars - Iron
Jupiter - Tin
Saturn - Lead

The Alchemist, by Pieter Brueghel The Elder, 1558.

An enormous amount of nonsense resulted from these supposed relationships, as well as an enormous amount of time wasted by otherwise keen minds such as Newton's. While the goal of alchemy was noble enough—the creation of the Philosopher's Stone which could cure all ills and transmute base metals to gold and silver (although Newton declared that the true seeker wasn't concerned with that aspect)—the pseudo-science became so immersed in magic, astrology, and its own over-inflated self-image, that it is difficult to comprehend how the same man who discovered gravity and calculus could have become so obsessed by it.

It is estimated that Newton spent twenty-five years in the study of alchemy, often going many days without food or sleep while conducting experiments. At the time of his death, his library contained hundreds of volumes on the subject and he, himself, wrote over half a million words concerning his research. As to the results of these years of intensive study, Keynes concluded that they were, "Interesting, but not useful, wholly magical and wholly devoid of scientific value."

Fortunately, this bad apple of Newton's career is overshadowed by his many accomplishments, but more than one scientist has no doubt lamented the fact that such a brilliant mind devoted so much time to the fruitless pursuit of alchemy.

Dephlogistocation

While the science of chemistry does owe a debt to alchemy for starting it on its path, the many misconceptions of alchemy are also responsible for blocking that path with some major obstacles. Take phlogiston, for example, that mysterious substance that was thought to be within all combustible substances and metals that corroded.

The ancient concept of the four elements—earth, air, water, and fire—was altered in 1667 by the German alchemist Johann Joachim Becher. He removed fire and air from the list and substituted three different forms of earth, one of which, *terra pinguis,* was supposedly an essential part of the composition of combustible materials, and was released when these materials were burned. In 1703, a professor Georg Ernst Stahl altered Becher's theory and renamed *terra pinguis* phlogiston, from the Greek word for fire.

To "prove" the existence of phlogiston, many good experiments were conducted with bad interpretations. For example, when wood

 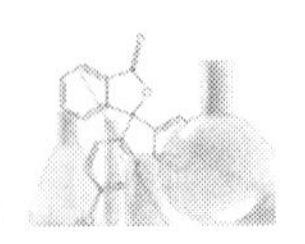

was burned, the total weight of the ash was much less than the original weight of the log, so it was concluded that the weight difference was the result of the phlogiston in the wood being released into the air during combustion. Also, when burning wood was closed in a sealed container, the fire went out, supposedly due to the "fact" that air could only hold so much released phlogiston. When air was full of phlogiston, it was "phlogisticated air" and when a substance was burned, it was "dephlogisticated."

Confused yet?

With all this multi-phlogistication research being conducted, many budding chemists spent years, and even entire careers, wasting time studying the mysterious and elusive substance. The theory should have been completely dephlogisticated after chemist Antoine Lavoisier demonstrated in 1770 that oxygen was the element involved in all combustion. Remarkably, however, the phlogiston theory was still embraced by some scientists until the early 1800s—most notably by Joseph Priestley, the man credited with the discovery of oxygen!

> Every attempt to employ mathematical methods in the study of chemical questions must be considered profoundly irrational and contrary to the spirit of chemistry...if mathematical analysis should ever hold a prominent place in chemistry -- an aberration which is happily almost impossible -- it would occasion a rapid and widespread degeneration of that science.
>
> Auguste Comte, 1798–1857

Who Needs Scientists and Chemists, Anyway?

Antoine Lavoisier was a born into a wealthy family in 1743, and rather than live the useless and dissolute life of a nobleman, he studied science and mathematics and became a brilliant chemist. His experiments and contributions to the field were enormous, and his work essentially set chemistry on the path of modern science.

Unfortunately, he was born in France, and remained there during the French Revolution.

Lavoisier had several strikes against him. For starters, he was intelligent and educated—and if there's anything that ignorant and violent mobs despise, it's people who aren't ignorant and violent. Lavoisier was also rich, of noble birth, and was associated with the Ferme Générale, a private tax collection agency. Perhaps his fate had really been sealed years earlier, however, when he snubbed one Jean-Paul Marat, who had created a useless invention which measured nothing.

Marat would become a leading figure of the revolution and denounced Lavoisier as a traitor. Displaying the same scientific knowledge that led to his useless invention, Marat included in his charges against the chemist, that he had stopped the circulation of air through Paris! Of course, if Lavoisier had such powers over nature, he would have summoned a flood to sweep the murderous scum out of Paris.

Marat was murdered in his bathtub in 1793, but the accusations of traitor led Lavoisier to trial in 1794. An appeal was made to save the brilliant chemist's life so he could continue his valuable work, which could be of benefit to all mankind, but a judge interrupted the appeal with a statement that will go down in the annals of Bad Science: "The Republic needs neither scientists nor chemists; the course of justice can not be delayed."

Antoine Lavoisier, one of the founders of modern chemistry, was sent to the guillotine on May 8. Lagrange, the mathematician, aptly summed up the senseless murder when he stated, "It took them only an instant to cut off his head, but France may not produce another such head in a century."

In a sickening twist to the story, a year and a half after his head was cut off, the French government, in its infinite wisdom, finally decided Lavoisier was innocent after all. His property was returned to his wife without an apology, just the single line, "To the widow of Lavoisier, who was falsely convicted."

There's a saying about being a day late and a dollar short, but in this case, it was a year and a half late and a head short. It is sad to think what other discoveries Lavoisier could have made had he

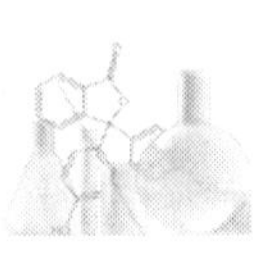

lived. But we will never know, thanks to the bloodthirsty people of France, and a judge who saw no use for scientists and chemists.

An engraving by Madame Lavoisier of her husband conducting an experiment on respiration. She is taking notes on the right.

> Science is a wonderful thing if one does not have to earn one's living at it.
> Albert Einstein

> Research is what I'm doing when I don't know what I'm doing.
> Wernher Von Braun

Birth, Contraception, and Sex

It's The Thought That Counts

One of the defining characteristics of the Middle Ages was the intense religious fervor which pervaded every aspect of life, as brief as those lives may have been. Infant mortality was high, as was the mortality rate for mothers during delivery. Surprisingly, the intense pain and suffering of women during that era was not always looked upon with sympathetic eyes—after all, it was Eve who had committed the original sin for which all women would have to forever pay the price. (And that seemed fair?)

As girls entered life with one big strike against them (i.e., being female), they also had to face a host of diseases, malnutrition, and general filth. Perhaps we should not wonder at the high mortality rates, but instead marvel at the fact that so many managed to survive. However, it should be kept in mind that survival of the body was not viewed as being all that important. Preservation of the soul was all that ultimately mattered.

As a result, difficult births that threatened the life of the baby demanded immediate attention—not for the purpose of saving the child, but to baptize it while it was still inside the mother. Rather than developing instruments and techniques that might have saved thousands of lives, an intrauterine baptismal syringe was developed.

 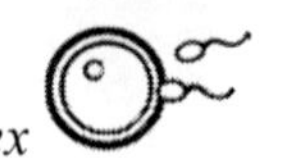

Once the elaborate device was inserted, the proper words spoken, and the water squirted onto the baby, the clergymen and doctors felt their work was complete. They would pack up their syringes with great satisfaction and leave the suffering mother and newly-baptized child to die at their leisure. (As the syringes weren't sterilized, even if the mother survived the delivery, an ensuing infection might finish her off later.)

Oh well, it's the thought that counts.

It would not be until the Renaissance that a more enlightened and compassionate view was taken on childbearing. It has been a long and painful road to the present state of health care, but considering it took until 1996 for a Presidential order to allow women to remain in the hospital more than 24 hours after giving birth, it is clear the road has not yet ended.

The Birth of an Anesthetic

In 1847, Scottish obstetrician James Simpson was experimenting with different chemicals in search of a suitable anesthetic. He and several colleagues would take a whiff of a substance and note any reactions. One evening, they all inhaled chloroform, and didn't wake up until the next morning! His next trial subject was his niece, whom he also rendered unconscious. Convinced of the chemical's capabilities, Simpson then began to use chloroform to alleviate the terrible pains of childbirth.

However, not everyone hailed him as a hero. Many clergymen condemned Simpson because women were *supposed to experience severe pain,* as the scriptures ordained that, "In sorrow thou shalt bring forth children." Simpson countered by saying that God had put Adam into a deep sleep to bring forth Eve. The critics were even further silenced when Queen Victoria used chloroform when she gave birth to Prince Leopold in 1853.

When Dr. James Simpson knocked himself out with chloroform, he knew he had found a viable anesthetic.

Burning Curiosity

For over a thousand years in Europe, male doctors were basically forbidden to attend a childbirth. Not that many would have wanted to anyway, as it was often looked upon as something beneath their station. Physicians who did attempt to enter the closed world of the midwife were severely criticized—or worse.

In Hamburg, Germany in 1522, a curious Dr. Wertt wanted to study a woman as she gave birth. Since no men were allowed at such an event, the doctor dressed himself as a woman. The disguise was discovered, and for this minor and well-intentioned transgression, Dr. Wertt was burned at the stake! (And interns think they have it rough.)

In rare instances, a physician would be called in if the baby was deemed undeliverable due to its position. The physician would be called not to try to correct the position, but to use his instruments to pull the baby apart and extract the pieces. Such appalling practices were still in use in the 16th century when the noted surgeon Ambrose Paré reintroduced the ancient technique of podalic version. The technique was simple and very effective—reach inside the mother, grab the baby by the feet and adjust its position so it could be delivered.

However, even as this technique became widespread over the centuries, overly-modest women did not make the job any easier. Often, the doctor was required to tie a sheet around his neck and place it over the woman so he could use his hands, but not his eyes. Strong objections to men being present at childbirth persisted into the 19th century, when it was even suggested that it was dangerous and potentially sinful to have a man attend to a woman in such a condition.

In actuality, given the nature of labor, what safer time is there for a man and woman to be alone together?

Clamping Down on a Family Secret

Catherine de' Medici (1519-1589) was not a happy ruler. Born to the powerful Italian de' Medici family, she was married to King

Henry II of France. Unable to provide an heir for the first decade of their marriage, she was much maligned by the people of her adopted country (although in the next thirteen years she would have ten children, three of whom would become kings). In addition to her ill treatment, she was further displeased by her husband's open affair with his father's ex-mistress, Diane de Poitiers.

When the opportunity arose, Catherine decided to take out her frustrations the old-fashioned way, murder—the murder of French Protestants (Huguenots) to be more specific. To escape the massacres, many Protestants fled to England, including the physician William Chambellan, who would later Anglicize his name to Chamberlen. The doctor's two sons, Peter and Peter (no, this is not a misprint) followed in their father's medical footsteps. Their claim to fame was the invention of obstetrical forceps to manipulate and extract babies during difficult deliveries.

The Peters' invention could have improved the lot of expectant mothers throughout Europe in a time when help was desperately needed. However, the forceps did not leave English shores, not because of a spiteful government, but because the Peters kept it a secret. The reason was simple; if *everyone* had them, *they* wouldn't make as much money. Repeatedly attempting to form a midwifery monopoly, the brothers ruled with iron forceps.

Passing on the family secret to the next generation, the younger Peter's son, Peter, (you knew that was coming) kept the family business going, still refusing to share this important knowledge with the world. They even went so far as to clear the delivery room and blindfold the woman in labor to keep the forceps a secret. The third generation of Chamberlens also continued to bring English babies into the world via the secret forceps. But as things got a little politically sticky in England during the second half of the 17th century, Hugh Chamberlen (caught you this time) decided to return to France where things were a bit safer.

Upon his arrival in Paris, Hugh boldly declared that he possessed the ability to deliver the most difficult cases. This time, he was willing to share his knowledge, for a price. Before the hefty sum was paid, the French obstetrician, Mauriceau, challenged Hugh to prove his claims. The unlucky guinea pig in the experiment was a pregnant dwarf whose body had been deformed by rickets. No conventional techniques were able to deliver her child, so Hugh and his forceps eagerly went to work. After three bloody hours of unsuccessfully digging around with his forceps, the experiment was over and the poor woman died.

Hugh returned to England, but later sold his secret in the Netherlands, where it was also kept under wraps. Finally, during the life of Hugh's son, Hugh (what a surprise), the forceps where made public.

It is sad to think that for a century, greed prevented the alleviation of so much suffering and death. It is even sadder to think how little things have changed.

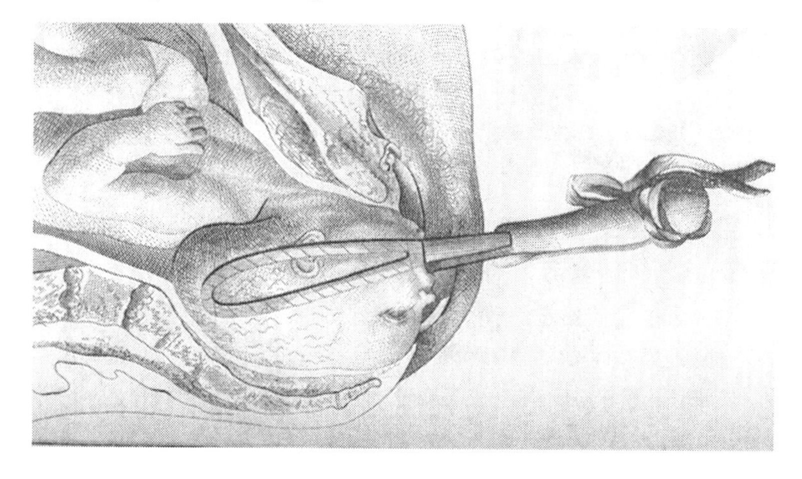

An illustration of the use of obstetrical forceps in 1792.

Take Two Aspirin and Get Burned at the Stake

The intense pain of childbirth can last for many hours, even days. Few women today would even think of enduring labor without some type of painkiller. However, there was a time when you could be killed for trying to alleviate the pain.

In Scotland in 1591, a "lady of rank," Eufame MacLayne, was in labor and about to deliver twins. She asked her midwife, Agnes Sampson, for something to help relieve the pain. Agnes obliged with some herbs. It was to prove to be a fatal mistake for both of them.

Unfortunately, Agnes Sampson was suspected of being a witch, and by treating Eufame, she was also implicated. The other problem was simply that a woman was not supposed to seek relief from pain during childbirth, because it was written in the Bible that God had specifically said to Eve, "I will greatly increase your pains in childbearing; with pain you will give birth to children." (Genesis 3:16). Therefore, church authorities determined that if a woman avoided pain during childbirth, she was going against God's will!

For these two grievous crimes of associating with an alleged witch and seeking to relieve pain during the delivery of twins, Eufame MacLayne was condemned to death. She was tied to a pole on Castle Hill in Edinburgh and burned alive. Agnes Sampson was also executed.

Even into the middle of the nineteenth century there was some resistance to the idea of easing labor pains when Scottish physician James Simpson began using chloroform.

If only men could get pregnant—the history of the world would be so different as to be unrecognizable today.

In Europe, midwifery came under the scrutiny and control of the church as it was feared that witches would intentionally cause the death of babies as offerings to the devil. Many midwives used herbs to ease the pain of childbirth, casting further suspicion that these women were witches casting spells. From the mid-15th to the mid-18th century, it has been estimated *that over 25,000 midwives were executed for witchcraft.*

Breeding Like a Rabbit

Those who have very large families are often referred to as people who "breed like rabbits," but one woman in England in the 18th century took things a bit too literally. She actually claimed *to give birth to rabbits*!

By all accounts, Mary Toft was an ignorant and uneducated woman. Born in 1701, she lived in the town of Godalming, married, and had several children. In 1726, Mary was pregnant again, but miscarried after claiming to have chased after a rabbit in the field where she was working. She became obsessed with rabbits, kept dreaming about rabbits, and craved their meat.

Two weeks later, Dr. John Howard was called upon as Mary apparently had gone into labor, despite the recent miscarriage. Lo and behold, Mary did deliver something—several parts of small rabbits! In the following days, more bits of rabbits emerged from Mary's birth canal, which was beginning to resemble something of a Rabbit Highway. Amazed, Dr. Howard spread the word to colleagues, inviting them to witness the miraculous rabbit births. The following article about Mary Toft appeared in the *Weekly Journal*, November 19, 1726:

From Guildford comes a strange but well-attested Piece of News. That a poor Woman who lives at Godalmin, near that Town, was about a Month past delivered by Mr John Howard, an Eminent Surgeon and Man-Midwife, of a creature resembling a Rabbit but whose Heart and Lungs grew without its Belly, about 14 Days since she was delivered by the same Person, of a perfect Rabbit: and in a few Days after of 4 more; and on Friday, Saturday, Sunday, the 4th, 5th, and 6th instant, of one in each day: in all nine, they died all in bringing into the World. The woman hath made Oath, that two Months ago, being working in a Field with other Women, they put up a Rabbit, who running from them, they pursued it, but to no Purpose: This created in her such a Longing to it, that she (being with Child) was taken ill and miscarried, and from that Time she hath not been able to avoid thinking of Rabbits. People after all, differ much in their Opinion about this Matter, some looking upon them as great Curiosities, fit to be presented to the Royal Society, etc. others are angry at the Account, and say, that if it be a Fact, a Veil should be drawn over it, as an Imperfection in human Nature.

Eventually, the king's own physicians came to see Mary, and they, too, claimed that she did indeed give birth to rabbits—at least parts of them. This seemed plausible due to a common belief that whatever a woman thought or dreamt about could affect her unborn child. As Mary had a thing about rabbits, it apparently made sense to some doctors that her body was somehow able to produce actual rabbit flesh.

 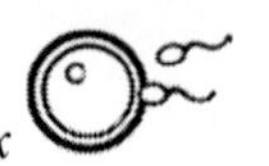

Despite the fact that only pieces of rabbits had come out of Mary Toft, this illustration shows lots of happy, hopping bunnies being born.

However, more skeptical witnesses examining her bizarre offspring also saw what looked to be various parts of cats, and a hog's bladder that still smelled of urine—despite the fact that Mary never mentioned being obsessed by cats or pig's urinary systems. Upon dissecting one of the rabbit's torsos, the feces in the intestines were found to contain bits of hay and seeds—substances that were unlikely to be consumed in the womb.

As England was in an uproar about Mary Toft giving birth to rabbits—to the point where the consumption of rabbit meat had all but ceased—it was clear something needed to be done. Mary was brought to London on November 29th, and large crowds gathered to catch a glimpse of her. She was kept under watch night and day, and lo and behold, there were no more rabbit babies, cat's legs, hog's bladders or any other animal parts delivered.

Mary wouldn't confess to the hoax, even after someone admitted supplying her family with small rabbits. Then a prominent London physician, Sir Richard Manningham, threatened to perform surgery on her uterus to discover why she was allegedly producing rabbits.

As 18th century surgery was just a slightly more refined type of butchery, Mary finally told the truth.

Claiming she was put up to the hoax by relatives and friends, she had chosen to stuff all sorts of dead animal parts inside of her in order to become famous and make money, thinking that the king would grant her a pension for her remarkable ability. Instead, she was granted a prison cell for being a "Cheat and Imposture in pretending to have brought forth 17 prætернatural Rabbits." Several of the doctors involved were disgraced, and the entire medical profession suffered quite a blow to its image.

Mary Toft, portrayed, of course, with a rabbit in her lap.

Satirists had a field day, including scathing illustrations by William Hogarth, and a poem written by Alexander Pope and William Pulteney with the following verse that stated a universal truth:

Most true it is, I dare to say,
E'er since the Days of Eve,
The weakest Woman sometimes may
The wisest Man deceive.

Mary Toft was released from prison after four months, and returned to obscurity, passing away in 1763. Despite the abuse of her reproductive parts, she actually did give birth one more time after her release, to a very human baby boy. Records do not indicate whether this child was named Hopalong.

Inconceivable Conception

In 1750, Abraham Johnson sent a report to the British Royal Society entitled *Lucina Sine Concubitu: A Treatise Humbly*

 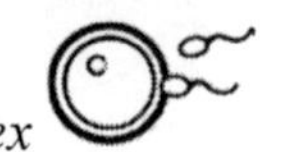

Addressed to The Royal Society. The society received many papers, but this one surely caught their eye, as the translation of the title is "Pregnancy Without Intercourse."

The title page further explains the content: "In which is proved, by most Incontestable Evidence, drawn from Reason and Practice, that a Woman may conceive and be brought to bed, without any Commerce with Man."

Within the text, the author states that this is possible, due to "floating animalcula," or microscopic human forms floating about in the air, just looking for a sleeping woman to impregnate. These animacula were observed by using "a wonderful, cylindrical, catoptrical, rotundo-concavo-convex machine."

Lucina ſine Concubitu.

A

TREATISE

Humbly addreſſed to the

ROYAL SOCIETY;

IN WHICH

Is proved, by moſt Inconteſtable EVIDENCE, drawn from Reaſon and Practice, that a WOMAN may conceive, and be brought to Bed, without any Commerce with MAN.

Ore omnes verſa in Zephyros ſtant rupibus altis,
Exceptantque leves auras, et ſæpe ſine ullis
Conjugiis vento gravidæ (mirabile dictu)
Saxa per et ſcopulos et depreſſas convalles
Diffugiunt, &c. VIRG. Georgic. iii.

Cur ego deſperem fieri ſine conjuge mater,
Et parere intacto, dummodo caſta, viro!
OVID, Faſt. v.

Or, as other Authors ſing,
The frolic Wind that breathes the Spring,
Zephyr with Aurora *playing,*
As he met her once a Maying,
Fill'd her with thee a Daughter fair,
So buxom, blithe, and debonnair.
MILTON's L'Allegro.

LONDON.
1750.

Going beyond the mere scientific importance of the discovery of these tiny human forms, Johnson says that this knowledge will help restore the honor of women who have some explaining to do, e.g., those wives who became pregnant while their husbands were away. He also came up with a method to prove his theory of animacula—stop all sex for one year by royal decree and then see how many women still get pregnant!

Where could such a bizarre idea originate? Did Johnson just pull his theory of floating animacula out of thin air? Actually, no, it was based upon a long-standing misconception of conception.

In the 17th century, when Antonie van Leeuwenhoek examined fresh sperm through his microscope, he thought he saw tiny, complete human forms within the head of the sperm. This "discovery" reinforced the idea that women were just incubators for the male "seed." Called homunculi—meaning little humans—the concept of "preformation" of a baby within the sperm spoke as much to the lowly perception of women as it did to the primitive level of scientific thought.

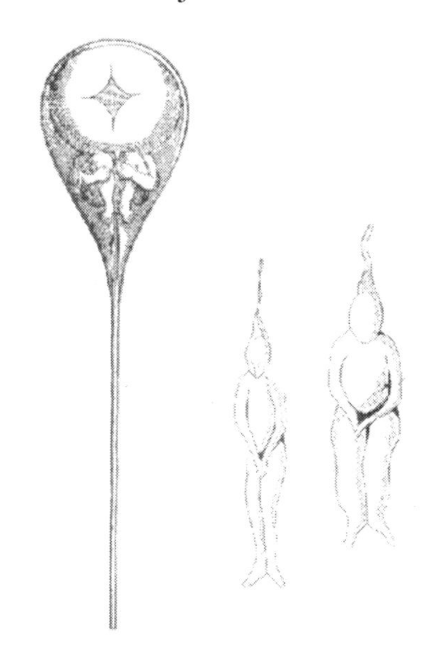
Homunculi in sperm as drawn by N. Hartsoecker in 1695.

For example, such a theory shouldn't have lasted more than a minute when considering the fact that children resemble their mothers as much as their fathers. However, to explain away this obvious refutation of the concept of homunculi, learned men concluded that the developing baby could "absorb" some of its mother's characteristics. (Another desperate and pathetic use of the age-old practice of molding the "facts" to fit your theory!) This theory persisted for generations, but not everyone agreed with it. In fact, some found it the perfect fodder for satire.

Lucina Sine Concubitu was actually just such a piece of satire, written by Sir John Hill, who led an interesting life. Hill was a botanist who compiled the massive 26-volumes of *The Vegetable System*. Obtaining a medical degree in Edinburgh, he was also an editor and author who managed to offend just about every major literary figure of his time.

 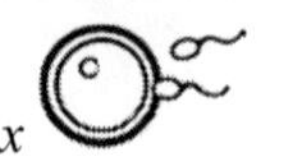

Hill expected that his work would give him entry into the British Royal Society, but he may have offended some scientists, as well, as his admission was denied. Perhaps this professional snubbing was, in part, motivation for targeting the society with his treatise on the nature of pregnancy without sex. Whatever the motivation, however, it stands as a great work of satire.

Of course, ignorance being what it is, some people actually *believed* it and promoted the idea of floating animacula. Among those proponents were no doubt some of those wives who had some explaining to do…

Back in the Saddle Again

The future didn't seem to be too bright for the nine children born to a poor saddle maker in Edinburgh in the mid-18th century, but one of those children was determined to make a name for himself, and in the most extraordinary way.

James Graham was born in 1745, and despite his family's circumstances, he attended medical school. While he left school before he got his degree, it didn't stop him from calling himself a doctor and attracting a clientele of the rich and famous. His specialty? Sex therapy, by means of employing the miraculous powers of electricity. In short, he helped couples get back in the saddle again.

Before embarking on that particular career path, however, he embarked on a ship to the United States. As the young "Dr." Graham was already pretending to be a physician, it was a small step to proclaiming himself a specialist. He traveled around the country for a while, but decided to set up his practice in Philadelphia as an eye specialist.

It was a fortuitous choice of cities, as Graham's own eyes were opened by the experiments going on at the time on electricity. While later in life he claimed to have learned the secrets of the powers of electricity from none other than Ben Franklin himself, Franklin wasn't even in Philadelphia when Graham was there. It is possible Graham had attended some electrical experiments conducted by one of Franklin's associates, but that doesn't make as good a story. And let's face it, if you are lying about everything else, why nitpick about something as trivial as being personally acquainted with one of the most famous men in history?

So, what did Dr. Graham, eye specialist, and Ben's new BFF, decide to do with electricity? Did he want to use it to cure hideous diseases and maladies that strike mankind? Not quite. The idea these electrical experiments inspired was this: Graham was "suddenly struck with the thought that the pleasure of the venereal act might be exalted or rendered more intense if performed under the glowing, accelerating and most genial influences of that Heaven-born, all-animating element or principle, the electrical or concocted fire."

[Author's Note: For those of limited vocabulary or imagination, the *Oxford English Dictionary* defines venereal (derived from Venus) as: "Of or pertaining to, associated or connected with sexual desire or intercourse."]

In other words, he planned to literally put a spark back in the lives of wealthy Europeans with, shall we say, flagging libidos.

[Author's Note: If I have to explain that you might as well skip ahead to the next story.]

In 1775, Graham returned to Europe and traveled about with his electrical "cures." He made some important connections with influential people, and he was emboldened by his success to open the magnificent Temple of Health in London in 1780. He spared no expense in the lavish décor, and even had scantily clad women, such as "Vestina, Goddess of Health," on display.

Visitors to the temple paid exorbitant entrance fees to attend Graham's lectures, listen to music, inhale the incense and perfumes that wafted through the building (early aromatherapy), examine electrical apparatus, and, of course, ogle the half-naked women. The showpiece, however, was unquestionably the Celestial Bed, a massive 12-foot by 9-foot tilting bed with mirrors on a suspended canopy, and charged with electricity generated by a man turning a crank in the next room. The bed was guaranteed to make the infertile fruitful and the impotent rise to the occasion—and stay risen for quite a while.

Graham wrote about his miraculous Celestial Bed, that "They all found that the pleasure was rendered not only infinitely more intense, but at the same time, infinitely more durable." And "When

 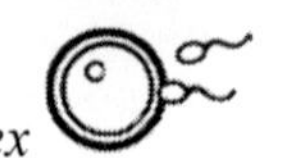

they were merry over a glass, they talked not as other men might have done, of the happy minute or of the critical moment - no! - they talked comparatively *of the critical hour*."

So, what did this 18th century version of Viagra cost? The happy—and rich—couples would shell out 50 pounds for a single night in the Celestial Bed—today's equivalent of thousands of dollars. Money came streaming into the Temple of Health and Graham's pockets. However, upkeep on the Temple was high, and when the novelty of his electric treatments waned, so did Graham's fortunes.

Forced into bankruptcy, Graham lost his Temple and Celestial Bed, and went back to Edinburgh in 1784. He then abandoned electricity in favor of dirt, and began promoting the miracle health effects of mud. He claimed that people could absorb all the nutrients they needed if they buried themselves up to their necks in mud, and he even went so far as to say he was immersed in dirt for two weeks with no food and felt fine afterward. Unfortunately for him, and fortunately for the public, this practice did not catch on. Neither did the religion Graham founded, or his habit of taking his clothes off in public to give to the poor (a practice for which he was arrested).

Despite Graham's belief that "earthbathing" was the key to a long life, he passed away at the age of only forty-nine, but what a half-century of life he had! Perhaps in retrospect with today's posh, expensive spas that offer mud baths and all manner of bizarre treatments, Graham was not a quack, just a man ahead of his time? Considering even now how many "little blue pills" and "male enhancement" products are on the market, wouldn't electrical stimulation still find buyers today?

Quack or not, it takes a gullible and desperate public to fall for such schemes, and if Graham was alive today and built another Celestial Bed, there would be a waiting list as big as the male ego.

Childbed Fever

While introducing new ideas is rarely easy, few scientists have had such a difficult career and tragic end as Ignaz Philipp Semmelweis. For starters, he was Hungarian, which in elitist Vienna in the 1840s placed one very low on the social ladder. Secondly, he was attempting to prove something that would cast the medical establishment in a very bad light. In fact, it was a fatal light.

The last known photograph of Semmelweis, c.1864.

Semmelweis maintained that the manner in which doctors conducted examinations of women was the cause of the extremely high death rate amongst new mothers. In other words, he was saying that even the most eminent physicians had the death of thousands of women on their dirty hands—an accusation that was obviously resented.

And let's face it, he was dealing with a segment of the population that wasn't too highly regarded—women, and poor women and prostitutes, at that. The all-male medical establishment looked down upon obstetrics and felt that delivery should be consigned to midwives. And if a lot of women died giving birth, well, so be it, that's their lot in life. One Professor Dietl so eloquently expressed this resigned attitude to a high mortality rate when he wrote, "The physician should be judged by the extent of his knowledge and not by the number of his cures. It is the investigator, not the healer, that is to be appreciated in the physician."

A remarkable statement, indeed! It's like the old joke about the operation being a success, but the patient died. Only this was no laughing matter. Why should a physician bother himself with such trivial things like cures and healing?

So, what was the problem that too many pregnant women faced? Childbed fever, otherwise known as puerperal fever, which was an infection that ravaged the bodies of women who had just given birth.

 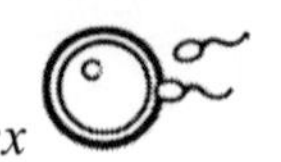

Most "lying-in" hospitals where expectant mothers went to deliver could anticipate mortality rates of anywhere from about ten percent to thirty percent. (In Lombardy, Italy in one year in the 1770s the town did have a perfect record—a perfectly awful record—as not one woman who gave birth in any of their hospitals survived!) Many women preferred to take their chances having a baby out in the streets rather than face such odds.

If you figure that one in every three women giving birth would die, you can see that the probability was very slim that any woman would live long enough to actually raise a few children. Of course, the physicians were curious as to the cause of these infections, and in their infinite wisdom came up with all kinds of ridiculous theories from changes in the weather, to strange "miasmas" (bad air), to unknown cosmic influences. The true and deadly answer, unfortunately, was literally at their fingertips.

The hospital in which Semmelweis worked was the Allgemeines Krankenhaus, or General Hospital. Checking the hospital's records, Semmelweis found that the mortality rate among women who had given birth there had started to climb after the facility became a teaching hospital in 1822. He also noticed that one of the wards was staffed only by midwives, and the mortality rate there was only two percent. What was the difference?

Medical students who worked in the high death-rate ward often conducted post mortem examinations on cadavers, and then went straight to conducting internal examinations on women. Consequently, at one moment a student could be elbow-deep in a pus-ridden corpse that had been decimated by infection, and just a few minutes later, those same unwashed hands could be probing inside a woman who had just given birth. And what a surprise—a couple of days later that same woman would be stiff and cold on a slab with another medical student cutting her up. And so on, and so on…

Although no one yet fully understood the concept of germs, Semmelweis saw a connection and ordered that everyone scrub their hands with a chlorine solution before examining patients. Soon after implementing the simple procedure, the death rate dropped from 13% to just over 2%. It was a brilliant success and should have meant that the lives of countless women would now be saved.

Unfortunately, Semmelweis then made the wrong move, or more accurately, made no move at all. Having already withstood considerable harassment from Viennese physicians, he refused to publish his findings. Some colleagues did try to get the word out, but it wasn't enough. When Semmelweis finally relented and did start publishing, the general reaction was just as he had expected—severe criticism.

While many argued that his findings lacked any scientific basis (which they did to some extent as germs were as yet unknown), physicians should have let the results speak for themselves. Instead, these doctors were insulted by the allegations that their ignorance was killing their patients. Remarkably, some doctors even complained that it would simply be *too much work* if they had to wash their hands *every* time they were going to perform an examination. It was obviously much easier *for them* to just let the women die.

There were those, however, who did see the light. Unfortunately though, for one German doctor the revelation came too late. He had recently delivered his niece's baby—without washing his hands—and she subsequently died of childbed fever. Overcome with guilt, the doctor committed suicide.

Semmelweis also struggled with intense feelings of guilt, knowing that he, too, had unwittingly caused the death of so many young women, but he did continue the fight. In 1861, he published his life's work in combating childbed fever in *Die Ätiologie, der Begriff und die Prophylaxis des Kindbettfiebers.* It was not enthusiastically received by the medical community, and the book also suffered from being almost obsessively detailed, as well as containing many personal attacks against Semmelweis' critics.

Legend has it that so many years of struggling against ignorance and prejudice took its toll on the emotional and sensitive Semmelweis, resulting in a nervous breakdown in 1865. He was committed to an insane asylum, where ironically, he died two weeks later from an infection that had been transmitted from a cadaver he had dissected before entering the asylum. A bad enough end to be sure, but the truth is even worse.

In 1963, Semmelweis' remains were disinterred and an autopsy was performed. Documents concerning his life were also examined, and a very different picture of Semmelweis' last years emerged. It

 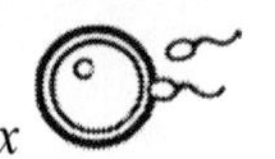

appeared as though he was most likely suffering from early onset of Alzheimers (he was only 47 when he died). He was committed against his will, and his remains indicated that asylum guards had severely beaten him. He died as a result of the untreated injuries, and possible infection, from the beating. Thus was the cruel end for a man who devoted his life to trying to save the lives of others.

In an obnoxious twist of irony in 1891, the people of Hungary—who once derided Sememlweis just as much as the Austrians—now realized that he had been right and wanted to bring their heroic countryman's body back home. The people of Vienna—who made Semmelweis' life hell—wanted to keep "their" hero's remains in Vienna. The Hungarians prevailed, and he was giving a proper burial and a statue, all just a little too late.

Granted, Semmelweis made mistakes and wasn't the greatest communicator, but both his cause and conviction made him the ultimate victor—even when the world had all but washed their hands of him…

A Stitch in Time

They say a stitch in time saves nine, but will wire stitches in your penis save you from the evils of masturbation? Gentlemen, hold on to your family jewels because this is going to get ugly…

Is there anyone who grew up in America who has not eaten Kellogg's Corn Flakes, or at least heard of their many breakfast cereals? From Frosted Flakes to Fruit Loops, millions of children and adults start the day with a crunchy bowl of some type of cereal. But what on earth, you are probably asking, does a corn flake have to do with mutilating your body to prevent you from masturbating? Just this—both ideas sprang from the same fertile and bizarre mind.

Dr. J.H. Kellogg was something of a health fanatic. He correctly believed that many diseases arose from improper diets. He therefore advocated vegetarian food, and thus invented the wholesome corn flake. He also maintained that the colon was the body's main trouble spot and was a great advocate of frequent enemas to keep your pipes clean.

However, above and beyond nutrition was a far greater danger lurking in the minds and pants of every boy and girl—the *disease* of masturbation, which, according to Kellogg, had profound "moral considerations" and could have the most dire "consequences to health of mind and body."

In 1892, Kellogg was the medical director of the Battle Creek Sanitarium in Michigan, and he published a book that included how to combat the dreaded practice of "self-abuse." In *Plain Facts for Young and Old: Embracing the Natural History and Hygiene of Organic Life*, Kellogg describes several clever ideas to thwart young boys' evil intents. For starters, "Covering the organs with a cage has been practiced with entire success."

With a cage!? Where could you buy such unique items, or did you have to have them custom made at your local blacksmith?

Another lovely idea was circumcision—but the key part of this strategy was that "the operation be performed by a surgeon *without administering an anesthetic*, as the brief pain will have a salutary effect on the mind." And if this young boy was not traumatized enough by this blatant act of torture for the sake of his alleged virtue, Kellogg compassionately adds that the bonus feature of circumcision is that, "The soreness which continues for several weeks" will "further interrupt" the desire for self-abuse.

But what if you're worried that circumcision won't work, that masturbation may continue after the soreness has ceased? Kellogg had an even more diabolical solution, although he admits (probably with regret) that he did not think of it himself.

"Through the courtesy of Dr. Archibald, Superintendent of the Iowa Asylum for Feeble-Minded Children," came the following, which is best described in Kellogg's own words.

"It consists in the application of one or more silver sutures in such a way as to prevent erection...the foreskin is drawn forward over the glans, and the needle to which the wire is attached is passed through from one side to the other. After drawing the wire through, the ends are twisted together, and cut off close."

The delightful result of having a needle jabbed into your penis and leaving in "one or more" metal stitches?

"It is now impossible for an erection to occur, and the slight irritation thus produced acts as a most powerful means of overcoming the disposition to resort to the practice."

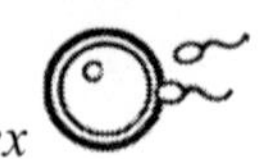

(Okay, gentlemen, take a moment now to breathe…Close your eyes and go to your happy place…If you're still feeling woozy, feel free to lean over and place your head between your knees for a few moments…Feeling better now? Good, let's continue.)

Of course, girls presented a more delicate problem. How best to treat this awful habit in young and impressionable females? Kellogg proudly declared that he had "found that the application of pure carbolic acid to the clitoris" was "an excellent means of allaying the abnormal excitement."

Dr. J.H. Kellogg, whose hobbies included enemas, inventing breakfast cereals, and sexually mutilating children.

Carbolic acid, also known as phenol, is used as an antiseptic. However, even in diluted forms it can cause skin irritation. Concentrated or "pure" carbolic acid can cause severe chemical burns. Can you imagine the result of this acid on the tender flesh of a young girl's private parts!

But why stop at merely burning the clitoris when you can completely remove it surgically? As early as 1858, Dr. Isaac Baker Brown wrote about this procedure, especially for those who were considered to be mentally ill. Dr. Kellogg did perform such surgery on a ten-year-old girl, but later actually had second thoughts—which was a little late for the poor girl he mutilated.

(Remarkably, such procedures are still performed in Africa today [to prevent girls from becoming promiscuous]—with razors and no anesthesia, but don't even get me started on the staggering ignorance and horror of that practice!)

So, tomorrow morning when you're crunching away on your corn flakes, think of Dr. Kellogg and his enemas. Or perhaps you should dig down deep into the cereal box to see if you have a special prize of a "Penis Home Suturing Kit," or a little bottle of carbolic acid.

Better yet, maybe you should just have some Post Grape Nuts…

Rule of 120

There's nothing quite so irritating as a group of men making decisions concerning women's rights, but such has been the long and unjust history of male-dominated societies around the world. From ancient civilizations where women weren't even considered as valuable as cattle, to present day countries where women aren't even allowed to show their faces in public, the rule of men has been rigid, harsh, and often cruel.

But surely in a country as modern and progressive as the United States in the 1960s, women had won the right to make decisions about their own reproductive systems…or not…

As late as 1969, the American College of Obstetricians and Gynecologists had an arbitrary and offensive "rule" in their manual of standards regarding when a woman was to be granted the privilege of voluntary surgical sterilization to free her from having any more children. Known as the "Rule of 120," it stated that a woman's age multiplied by the number of children to which she had given birth had to equal 120 in order for her to be eligible for the procedure. Apparently, the eminent physicians at the ACOG had, in their infinite male wisdom, decided that women had fertility quotas they were obligated to meet.

In other words, in 1968, an educated woman of forty who was running a successful business and had two grown children in college was not qualified to decide that she didn't want to have any more babies. According to the Rule of 120, at a score of only 80 she would most likely have been denied a tubal ligation. Similarly, a nineteen-year-old single mother of six children living in abject poverty would not have qualified, either.

In addition, this rule did not even take into account a woman's health, so the fact that an additional pregnancy might endanger the mother's life did not enter the equation.

It is easy to scoff at cultures where women were viewed to be no better than cattle, but let us not forget that in our very recent past, the American woman was still seen to be something of a breeding cow.

Egyptian Dick and Jane

An Ancient Tale of Contraception

See ancient Egyptian Dick make his move on Egyptian Jane. "Your eyes sparkle like moonlight on the Nile," Dick whispers in her ear as he takes her in his arms.

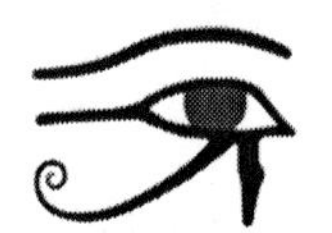

"Oh, Dick!" Egyptian Jane swoons, "Is that a papyrus in your loincloth, or are you just happy to see me?"

See Egyptian Dick and Jane start to go at it. Go Egyptian Dick. Go Egyptian Jane.

"Stop!" shouts Jane. "Stop, Dick, stop."

"But why, Jane" moans Dick. "Is it the sand again?"

"No, Dick, no. I must take precautions or else we will soon hear the pitter patter of little pyramid builder's feet around the house," Jane replies.

See Egyptian Jane prepare her contraceptive from a popular recipe. See her mix many ingredients and place them in a hollowed-out half of a small citrus fruit. See Egyptian Jane insert her contraceptive.

Carefully, Jane, carefully.

See Egyptian Dick and Jane start to go at it again. Go Egyptian Dick. Go Egyptian Jane.

"Stop!" shouts Dick. "Stop, Jane, stop!"

"But why, Dick?" Jane moans. "Is it the locusts again?"

See Egyptian Dick begin to gag. "In the name of Amon Ra, what is in that contraceptive?" Dick asks, holding his nose.

"It is a mix of the purest honey and herbs," Egyptian Jane replies, and then remembers the special ingredient. "Oh yes, the recipe also calls for crocodile dung."

See Egyptian Dick run. Run, Dick, run.

In the ancient world, linen, silk, oiled paper, and leather were used for condoms. Linen was still being used in London in the 18th century, but good condoms were expensive. This is what prompted the enterprising Miss Jenny to start gathering used condoms and washing them. She then sold the second-hand condoms at a discount.

Over the millennia there have been many interesting attempts at contraception:

- In ancient China, women drank hot mercury to prevent pregnancy. Of course, ingesting mercury is a great way to prevent all life.
- In ancient Rome, women would wear a leather pouch during sex to prevent conception. The pouch contained the left foot and liver of a cat. The only births this prevented were those of kittens.
- Even more bizarre than cat parts, were rabbit parts. In the Middle Ages, women wore amulets that consisted of the anuses of rabbits made into a wreath. Butt why?
- Another popular method that began in ancient times and was used into the 20th century was wood—a piece of wood inserted to cover the cervix and literally "block" the sperm. (Sure hope that wood was sanded to a smooth finish!)
- Last, but surely not least, in Canada, dried beaver testicles were ground into a tea or mixed with alcohol. Drinking this tea or a "beaver high ball" was supposed to prevent pregnancy. It did, but only for beavers.

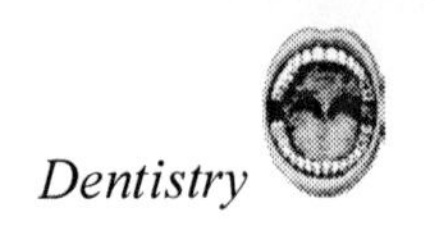

Dentistry

> The male has more teeth than the female in mankind, and sheep and goats, and swine. This has not been observed in other animals. Those persons which have the greatest number of teeth are the longest lived; those which have them widely separated, smaller, and more scattered, are generally more short lived.
>
> *Aristotle, 384-322 BC, History of Animals*

The Evil Tooth Worm

Dental caries, commonly called cavities, are not a recent occurrence in the history of man, despite our enormous intake of soda, bubblegum, and Twinkies. To be sure, our love of sugary snacks keeps many dentists snug and warm in their shiny new BMWs, but tooth decay has been tormenting hominids for at least a million years.

Today, we know that painful cavities are caused by bacteria that produce acids that eat into the tooth, but in ancient times, they blamed other living creatures—tooth worms. These worms were believed to chew holes in the teeth, then take up residence inside. It was when these worms were restless and started wriggling that the pain became intense. One of the first written accounts of this nasty worm was in an ancient Sumerian text, and the belief in these destructive critters was prevalent throughout the ancient world.

The worms were variously described as looking like maggots or eels, and being black, gray, red, or blue. Obviously, no one ever actually saw a tooth worm—for the simple reason they don't exist—so one can only wonder how common it was to have worms

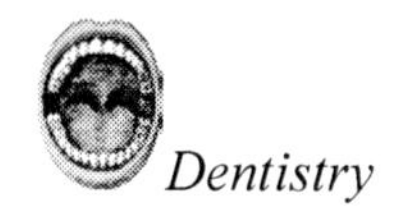

infesting your food—worms that could be mistaken for tooth worms as the food was being chewed.

Some thought the tooth's nerve was the worm, so it would be yanked out—but how anyone could remain conscious while their exposed nerve was being probed and pulled out I can't even begin to imagine! Some weight was given to this concept for the simple reason that once the nerve was removed the pain naturally ended in that tooth. Yanking out the entire tooth was also a way to rid yourself of the worm-induced toothache.

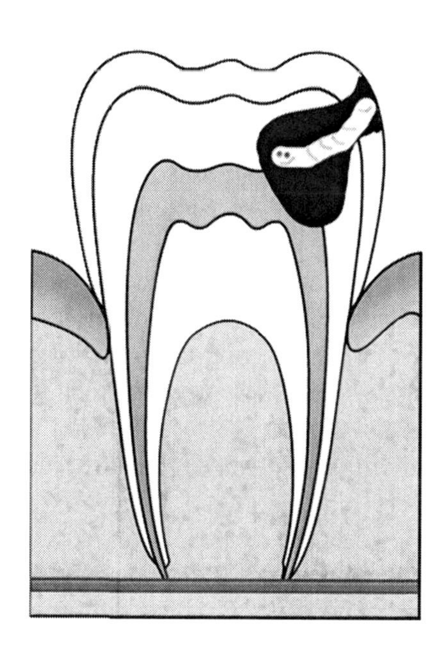

Other "remedies" included chanting spells and burning henbane seeds as a type of exorcism to drive away the evil-spirited worms. Henbane is a foul-smelling plant that in small doses could be used as a sedative, but could also cause hallucinations and death—both results of which may actually be preferable to a really killer toothache!

A medieval German text contained an alternate use of henbane as described in the following instructions: "If the worm hollows the teeth and eats the gums, you should take henbane oil and knead it with wax and form a candle out of it. Put the candle into a bowl filled with a little bit of water: As soon as the candle burns, you must hold your teeth over it and the worms will fall into the water."

Honey could be used as a type of bait to lure the worm out of the tooth. Another "cure" for a toothache was to wear a string of worms around your neck. When the worms died, you threw them into a fire and said a prayer. Another common practice was to wear the tooth of a dead person, with the teeth of a criminal who had been hung being particularly helpful in easing pain.

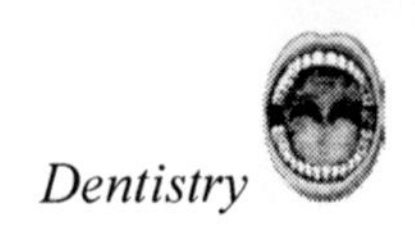

As the success rate for all these alleged remedies had to have been exactly nil, one wonders how these practices continued for as long as they did!

It wasn't until the 18th century that the widespread belief in tooth worms was finally replaced by a more scientific understanding of the mechanism of tooth decay. In a way, perhaps, that is unfortunate. At least when mankind felt that evil little worms were responsible for his rotting teeth, they had something to blame. Now, all we have is ourselves to blame for our lack of dental hygiene, as well as all the soda, bubblegum, and Twinkies, of course.

Paying *with* your Teeth

During the Victorian era, the sun never set on the British Empire. However, there were some regions were the sun never shone—namely, in the closed mouths of the self-conscious citizens whose teeth had rotted.

While bringing the art of subjugating native peoples to a new high, the British were nonetheless at the mercy of rebellious molars and bicuspids. Some resourceful ladies of the evening combated the problem by placing violet-scented crystals in their mouths to hide the rotting stench, but respectable people seen in daylight needed a way to mask the decay and loss of their pearly whites (or browns and greens, as the case may be).

The use of dentures was nothing new, as the earliest attempts go back to at least 700BC. Many materials were tried, including wood, porcelain, and animal teeth. Perhaps the most famous denture-wearer in history was George Washington, but contrary to popular belief, his choppers were not wooden; they were made of hippo ivory. Unfortunately, nothing quite looked like real teeth as real teeth did.

Hence, the brilliant Victorian denture solution—use real teeth. The dilemma, however, was where to find a constant supply of healthy teeth? One obvious tried-and-true solution—take advantage of the poor.

In a rather unsettling twist of irony, the starving and destitute of Victorian England would have their healthy teeth yanked out so they could sell them to buy food. Of course, if one continued this practice to its inevitable conclusion, not only would you have nothing more to sell, but it would be very difficult to eat anything even if you had it. Not that the wealthy recipient of the new human dentures lost any sleep over that thought.

There probably also wasn't a wink of sleep lost over the other harvesting techniques. For example, bringing a pair of pliers to a hanging. Or, if you didn't mind getting your hands dirty, grave robbing was a lucrative business to get into. Yet even these practices pale in comparison to the best get-rich-quick schemes in the denture trade—wrenching the teeth from the corpses on the battlefield.

The Battle of Waterloo, with its estimated 50,000 casualties, supplied a bumper crop to the enterprising entrepreneurs who raced onto the battlefields with sacks and pliers before the smoke had cleared.

So intense was this search for high quality teeth, that tooth hunter-gatherers even looked across the pond for supplies. After all, Americans had taken British soil, the least they could do now was cough up a few teeth in gratitude. Fortunately, the hunters were not disappointed, as the years 1861-65 provided a boom market.

The bloody battles of the American Civil War literally produced barrels of teeth that were shipped back to England so the idle rich could smile proudly when counting their money. And if a British gentleman was accused of lying through his teeth, he could respond in all honesty that if there were any lies being told, it was through someone else's teeth.

Fortunately, high quality artificial teeth have eliminated the need for such grisly practices, but even today, the high cost of dentures is often beyond the budget of the uninsured. However, pliers still remain quite affordable.

Rub Them Out

Just like today, an 1886 advertisement for *Sozodont* tooth polish blatantly played upon one of women's primary concerns, appearing

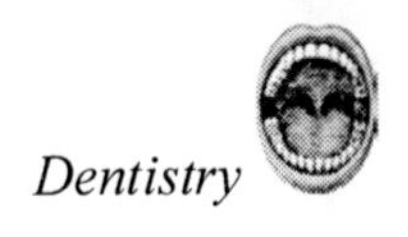

attractive to men. "Teeth of pearls and breath of roses are the winning charms of the fair sex," the ad began, and went on to describe that the ingredients were "purely vegetable," and therefore wouldn't "injure your teeth." Further claims were made that *Sozodont* "rendered enamel impervious to decay." This claim may, in a backwards way, have some validity. As the American Dental Association found that *Sozodont* "cut teeth like so much acid," it would only follow then, that with continued use, your teeth would quickly have no enamel left to decay.

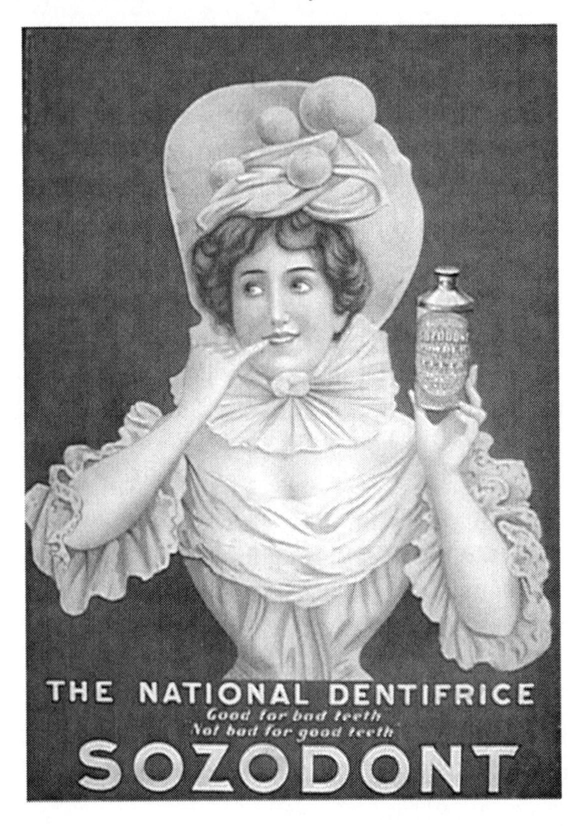

The American marketplace was filled with harmful whiteners and polishes well into the 20th century. During a 1935 study, *Tartaroff* was found to contain so much hydrochloric acid that with just one use, three percent of tooth enamel was brushed away. It doesn't take a rocket scientist to realize that with daily use, *Tartaroff* would best have been called *Tooth-B-Gone*.

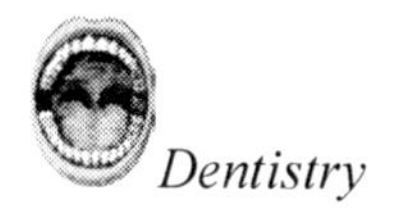

Hush, Little Baby

At the misty dawn of mankind, when the first tooth began breaking through the tender gums of our ancient ancestors' first baby, a universal experience for parents was set into motion for countless generations to come—sleep deprivation. How many eons-worth of fragile sleep has been shattered beneath the incessant cries of a teething baby? What parents would not welcome a remedy which quieted their little darling and allowed them to get a good night's rest?

In the early 1700s, the clever and enterprising British thought they had found such a remedy; not among the wild herbs and gardens of their native soil, but in the far-off lands of China. There they found an innocent looking flower, *Paparvera somniferum*, which yielded a milky juice from its seedpods which had the miraculous ability to calm even the crankiest of children.

So precious did this juice become to the British that they went to war over it in 1839 after China had forbidden further exportation of the flower. In brief, the British won the Opium Wars, allowing the prized opium poppy seedpods to flow freely into England once again. And, they got Hong Kong thrown into the deal, as well. However, the victorious British no doubt felt that mere territorial gains were nothing compared to winning the right to continue to drug themselves into a stupor.

Opium had become the panacea for whatever ailed a man or woman, so it only seemed natural that it would also be a great benefit to children. Concoctions made with this dangerous narcotic, such as *Dr. Godfrey's General Cordial* and *Dalby's Carminative,* were literally poured down the tiny throats of innocent children by their well-meaning parents.

These popular British concoctions quickly made their way to her colonies, as well. Anyone reading Benjamin Franklin's newspaper, *The Pennsylvania Gazette*, on June 26, 1732, would have seen the advertisement touting *Godfrey's Cordial* as a cure for "*Cholick, and all manner of Pains in the Bowels, Fluxes, Fevers, Small-Pox, Measles, Rheumatism, Coughs, Colds, and Restlessness in Men,*

 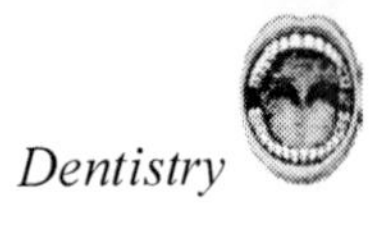

Women, and Children; and particularly for several Ailments incident to Child-Bearing Women and Relief of young Children in breeding their teeth." (But did it do windows?)

In the spirit of patriotism, the Revolutionary War set American pseudo-pharmacists to work on our own brands of the British mixtures, so we could poison our children without any foreign intervention. Each new mixture developed had different levels of opium (or one of its derivatives, morphine) so parents switching to a stronger brand could easily give their child a harmful overdose—if they hadn't already put them permanently to sleep with the first brand.

One of the most popular American teething remedies was *Mrs. Winslow's Soothing Syrup*, introduced in the 1830s. This syrup contained enough opium that, "Few children under the age of six months would not be poisoned to death, were they to take the syrup as directed," a California physician reported in 1872.

Unwitting parents turned their babies into addicts, overdosed them into comas, or tragically, killed them. It is impossible to tell how many children died from these opium remedies, but in 1776 in England, it was already estimated that thousands of children were inadvertently killed each year. Considering that consumption of opium remedies did not reach its peak until 1896 in America, the actual numbers of accidental deaths would no doubt be staggering.

Even more tragic was the fact that many children were given these drugs not because of pain or illness, but simply to keep them quiet. Opium became a poor mother's babysitter—by giving the child a large enough dose to put it in a stupor, the mother could go off to work to earn money to feed the child she was putting at deadly risk. Less scrupulous parents would even drug their children so they could go out on the town and have a good time without having to worry about what mischief Junior might be getting into.

Many fortunes were made at the expense of children's lives and health. Even after the dangers of opium and other narcotics began to be more fully understood in the 1870s, manufacturers continued to churn out their poisons into the 20th century.

Teething is still as painful today as it was a thousand generations ago, yet the simplest remedy remains the safest—earplugs for the parents.

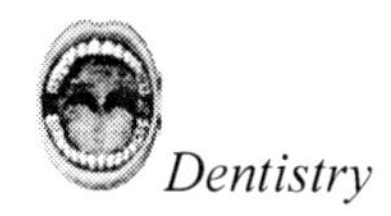

Washing your Mouth Out with *What*?

The first modern antiseptic mouthwash was *Odol*, which began being sold in 1893 by the German entrepeneur, Karl August Lingner, and it is still on the market today. *Listerine* had been introduced in 1880 by the Lambert Pharmaceutical Company, but it was originally sold as just a general antiseptic. Only when sales were waning after World War I, did the marketing team at Lambert come up with the idea to use *Listerine* to combat bad breath, sending sales sky high and creating one of the world's most recognizable products.

Some of the ancient remedies to promote fresh breath and avoid tooth decay would be unrecognizable today, and that is a *very* good thing. The Greeks rinsed their mouths with donkey milk, but that's nothing compared to what the Romans used.

Urine.

Human urine.

Preferably Portuguese urine, if you could get it.

Seriously.

While most people today would rather let every bicuspid and molar rot and fall out of their bleeding gums before they would take a swig of urine, the Romans had quite a trade developed in collecting and importing barrels of the golden liquid waste product. Portuguese urine was most sought-after, as it seemed that for some reason, their population's pee was more potent and had a higher acid content, which apparently kept it fresher during transportation. After all, who wants to rinse their mouths with stale, weak urine, when you can have fresh-tasting, robust Portuguese Pisswash.

As disgusting as this all is, there was actually some method to this madness—the ammonia in urine does have antiseptic properties. In fact, ammonia is still used in some mouthwashes today, although most brands now use alcohol, instead.

Of course, there was that whole bad taste issue with ammonia, so to try to mask the pungent ammonia flavor, over the centuries different ingredients were also added, such as honey and eucalyptus.

 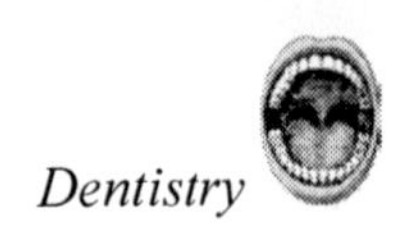

While these pleasant ingredients make perfect sense, others were quite baffling, such as lizard livers and ground up rodent heads!

So tomorrow morning when you roll out of bed and reach for that bottle of mouthwash you take for granted, just remember what people used to do to avoid bad breath and prevent tooth decay. And if your mouthwash happens to be a yellowish color, no one will blame you if you decide to switch brands.

Dim Bulbs

Comments on the incandescent light and Edison's improvements on the light bulb:

"When the Paris Exhibition closes, electric light will close with it and no more will be heard of it." Erasmus Wilson, Oxford professor, 1878

"...good enough for our transatlantic friends...but unworthy of the attention of practical or scientific men." British Parliamentary Committee, 1878

"Such startling announcements as these should be deprecated as being unworthy of science and mischievous to its true progress." Sir William Siemens, inventor, 1880

"Everyone acquainted with the subject will recognize it as a conspicuous failure." Henry Morton, president, Stevens Institute of Technology, 1880

Disconnected

"It's a great invention, but who would want to use it anyway?"
President Rutherford B. Hayes, commenting on a demonstration of Alexander Graham Bell's telephone, 1872

"This telephone has too many shortcomings to be considered as a means of communication. The device is of inherently no value to us."
Western Union, 1876

"The Americans have need of the telephone, but we do not. We have plenty of messenger boys."
Sir William Preece, British engineer and inventor, 1878

Geology, Paleontology, Archaeology
(and other things found in dirt)

Stone Blind

Be not arrogant in your knowledge, for wisdom is found even among the slave girls at their grinding stones.

- Ancient Egyptian saying

Arrogance has been the downfall of many great political and military figures. The academic world has also had its share of scholars convinced of the infallibility of their own minds. One such scholar was Dr. Johann Beringer, whose blindness to the truth proved that he was clearly a man who had his head firmly planted up his own assumptions.

Born in Germany in 1667, Johann Bartholomew Adam Beringer was the son of a prominent professor. Following in his father's footsteps, he became Senior Professor of the University of Würzburg and Chief Physician to the court of the local prince. When the fossilized bone of a mammoth was unearthed near Würzburg in 1710, it was Dr. Beringer who declared that it absolutely was not the leg of a giant man from the time of Noah, as the uneducated populace thought. In fact, the highly educated doctor asserted with authority, he was certain that it wasn't even a bone—it was a *lusus naturae*, or "prank of nature." Apparently, he believed that God got a kick out of making rocks that simply looked like plants and animals. As ridiculous as this seemed, what god-fearing man would dare question the Almighty's sense of humor?

Fortunately, there were a few men who did strongly disagree with the stupid theory, and unfortunately for Beringer, these men were also at the university. Professor of Geography and Algebra, J. Ignatz Roderick, and the university's librarian, Georg von Eckhart, both agreed with the theory held by Leonardo da Vinci two centuries earlier, that fossils were the result of plants and animals so ancient that their remains had turned to stone. However, this common sense explanation was far too common and far too sensible for a man who had spent his entire life in school, and Beringer treated everyone who disagreed with him with contempt.

Roderick and Eckhart decided it was time to teach the high and mighty professor a humbling lesson. They knew that Beringer had hired diggers to search the countryside for fossils, but for years had found nothing. Then one day in 1725, the diggers were going over ground previously searched and suddenly made a spectacular find—in fact, too sudden and too spectacular to be believable.

What they uncovered was a treasure-trove of stones bearing detailed carvings of plants, birds, insects, fish, mammals, and every manner of living creature. There were also "fossils" with scenes of comets, the Moon, and the sun! Impossible subject matter aside, they showed too many signs of having been recently carved. They were unmistakably, blatantly, too unbelievable to be true. However, because they conveniently supported his theory, Dr. Beringer believed them.

As the diggers continued for six months to uncover more stones, they came across some which were even more absurd—stones which contained inscriptions in Hebrew, Latin, and Arabic. Experts indicated that the inscriptions were all ancient names of God. Still not seeing the forest for the trees (into which he was about to run face-first), Beringer proudly announced to the world that here was finally the ultimate proof that fossils were the products of the hand of the Almighty, because God himself had signed some of his works of art!

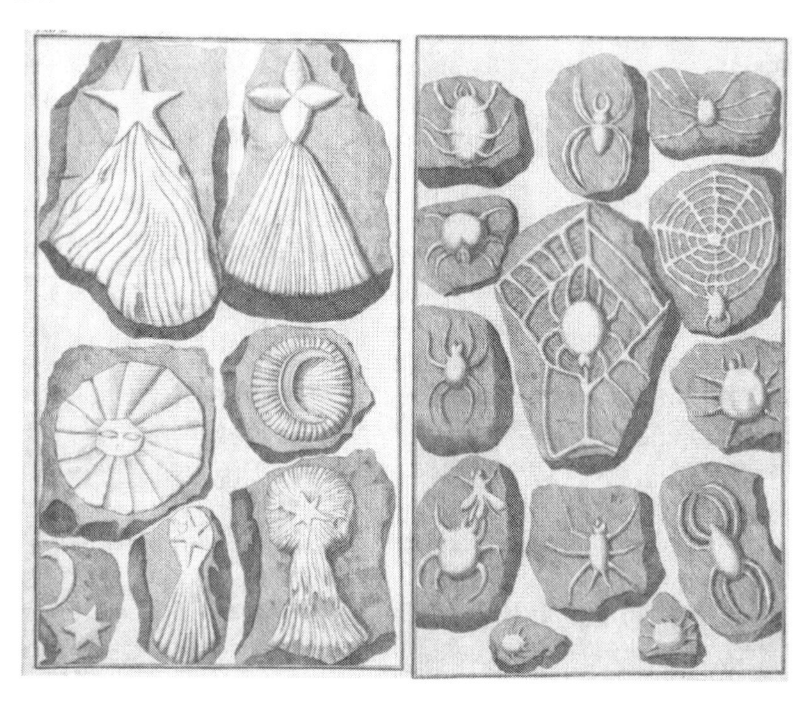

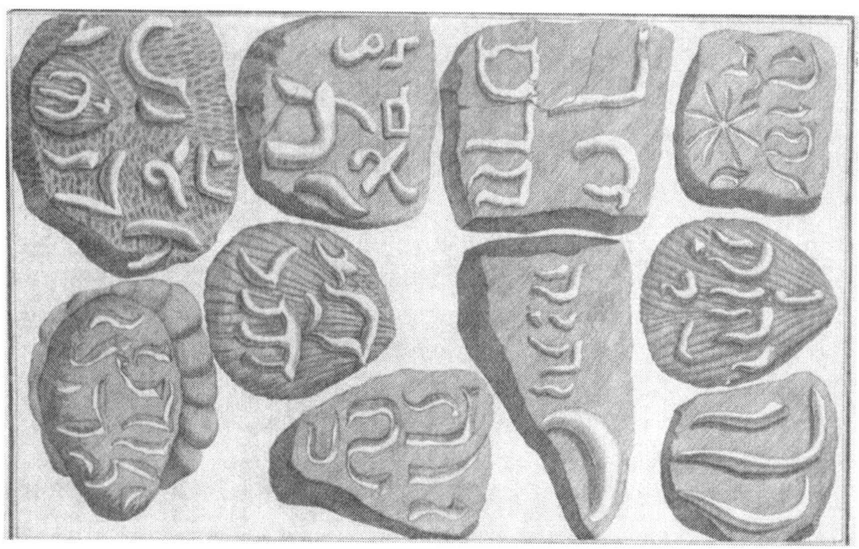

Beringer hired a famous artist to produce these drawings of the "fossils" of comets, spiders, and the names of God.

The actual stones now reside in museums.

Oblivious to the obvious, Beringer forged ahead. Roderick and Eckhart stated publicly that these stones were no doubt someone's idea of joke (although they conveniently neglected to identify themselves as the someones), but Beringer chalked up their reaction to academic jealousy, since *they* never found anything signed by God.

Then Roderick and Eckhart even admitted to carving stones which were sold to Beringer through one of the diggers. When word of this came out, an angered Dr. Beringer conceded that those few stones might be frauds, but all the thousands of others (which were carved in the same manner on the same type of stone) were absolutely authentic.

Had Beringer just quietly lectured at the university about his stones, history might have let him slip into peaceful oblivion. However, despite sage advice from his peers, Dr. Beringer invested a substantial sum of his own money to publish his finds in *Lithographiae Wirceburgenis.* (The rough translation of which means "Look how big my mouth is and watch how far I can put my foot into it.") Within the pages of the magnificently printed and bound book, Dr. Beringer actually states that the stones did bear "the unmistakable indications of the sculptor's knife," but adamantly insisted that sculptor was none other than God himself!

The book was an immediate sensation across Europe as other gullible scholars eagerly purchased copies. However, as the illustrious Dr. Johann Beringer approached the pinnacle of success, he quietly began trying to buy back all his books. It wasn't because he was so fond of them—he wanted to incinerate them.

Gossip spread like a book-burning bonfire—Dr. Beringer had just uncovered a stone that had his own name on it, and finally realized what a world-class idiot he had been. The hoax revealed, the books became highly-prized collectors' items and Dr. Beringer spent the rest of his life living with the eternal humiliation he had brought upon himself.

The hoaxsters, Roderick and Eckhart, didn't fare much better, as a court of inquiry revealed how they had brought shame to Würzburg's famous hometown boy. The incident incensed the townspeople who felt their city's name would become synonymous with stupidity, and Roderick prudently beat a hasty retreat. Eckhart kept his job at the library, but in the four short years left in his life, it is certain that university officials never discussed any career advancement plans with him.

Today, many of the stones are displayed in museums, and Beringer's book has been reprinted numerous times. On the bright side, not many authors can claim a three-hundred-year print run for their work!

As for the years of disgrace that the arrogant doctor endured, we can feel little pity for him. He carved his own name into the fossil record of Bad Science.

> That the automobile has practically reached the limit of its development is suggested by the fact that during the past year no improvements of a radical nature have been introduced.
>
> *Scientific American*, 1909

Skeletons in the closet

While most people dream of winning the lottery or having their mother-in-law move to a retirement community in a remote section of the Andes, paleontologists dream of finding that one bone which will change all the textbooks—preferably by the addition of their own name. Dr. Albert Koch made one such discovery in 1838, but due to some rather large skeletons in his closet, the only book in which his name might appear would be in a course of Con Artistry 101.

Koch's discovery in October of that year in Gasconade County, Missouri, was indeed a legitimate landmark in the history of science—bones of the extinct giant sloth, together with stone knives, axes, and a plentiful supply of ashes from the sloth roast. This was of great importance, because the prevailing scientific beliefs of the mid-nineteenth century held that mankind had inhabited the North American continent for no more than 2,000 years. Dr. Koch's revelation of the ancient barbeque should have pushed those estimates back at least 10,000 years. What could this enterprising German immigrant possibly have done to make the scientific community scoff at his claims?

For starters, he had bestowed the title of doctor upon himself. This slight indiscretion might have been overlooked, had it not been for a couple of other things that were to set the industry standards for being indiscreet—e.g., the massive skeletons of *Missourium* and *Hydrarchus*. It seems that the self-anointed doctor unearthed the spectacular bones of several mastodons along the banks of a river in Missouri in 1840. In themselves, the bones were among the finest ever found, and for their discovery Koch would have attained respect and prestige from the scientific community. However, he was more interested in obtaining what was in the public's wallets.

Taking bones from several of the giant mastodons, Koch constructed a skeleton of an animal whose size rivaled only that of his ego. As the crowning touch, he purposely affixed a set of the huge tusks on the top of the skull to make it appear as if the beast had fearsome horns. Christening his creative invention the *Missourium*, Koch took the mythical mastodon on a tour and made a fortune from trusting and eager audiences in the United States and Europe. Legitimate scientists denounced the fraud, but since when has reason been able to compete with entertainment value?

Remarkably, Koch's triumphant three-year European tour ended on an even more prosperous note—the British museum paid him a mammoth sum of money for the *Missourium* (although after purchasing it, the museum's scientists quickly rearranged the bones to their proper form).

Returning to America in 1844, Koch was anxious to dig up his next trick. In March of 1845, he discovered a treasure-trove of fossils in Alabama and spent the next few months piecing together a new creature. Playing upon the public's fear and fascination with sea serpents, Koch produced the 114-foot-long *Hydrarchus*, the alleged descendant of the Biblical Leviathan.

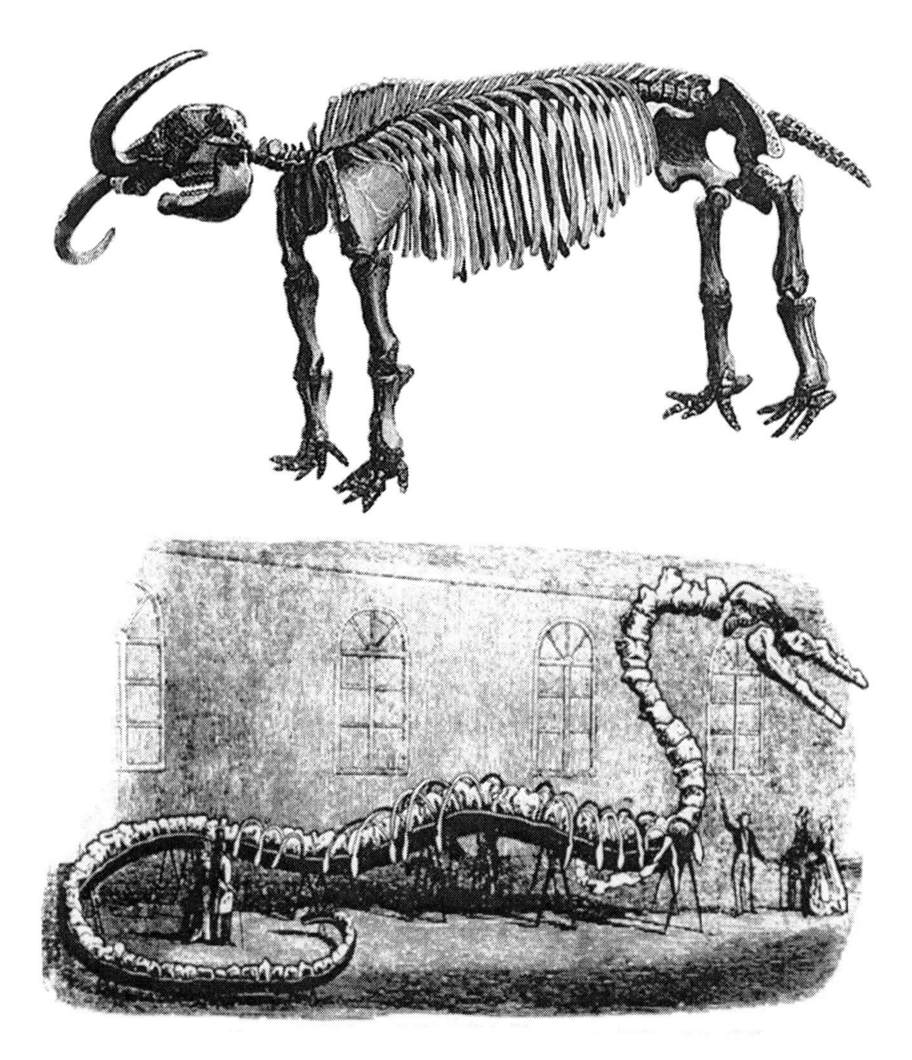

Missourium and Hydrarchus

The startling beast drew large crowds once again, despite evidence from the academic community that the creature had not sprung from the ancient, fertile seas, but merely from the fertile imagination of a marketing genius (which today, of course, would be a contradiction of terms). The British were less enthusiastic, however, since they had seen the smaller, de-horned version of *Missourium* and had realized they had been duped.

Koch's native Germany was far more accommodating and the Deutsche Marks rolled in. And since it's always easier to travel without a 114-foot-long skeleton, (although it seems people still try to ram objects of similar size into a plane's overhead compartments), Koch was doubly pleased to sell *Hydrarchus* to no less than King Frederick Wilhelm IV.

Koch was good for one more charade—a new *Hydrarchus* and yet another tour. But by the early 1850s, even he had enough of peddling frauds for fun and profit. Retiring to St. Louis, he tried for the remainder of his life to convince the scientific community that in 1838 he had indeed discovered evidence of mankind's early habitation of the continent. Yet for some strange reason, bogus horned beasts and mythic sea serpents wreak havoc with one's credibility, and no reputable scientist would believe his tales of sloth-bakes.

It wouldn't be until 1927, almost 90 years after Koch's find, that irrefutable proof of man's earlier habitation of the continent was finally uncovered by J. D. Figgins in New Mexico. However, even the purely legitimate Figgins was, at first, subject to the scorn of the scientific community, that body of learned individuals who often regard change to be a four-letter word. (See what a real Ph.D. does to you?)

The Cardiff Giant

George Hull didn't do so well with his tobacco farm in Binghamton, NY, so like many men he decided to head west to seek his fortune. In 1868, he was visiting his sister in Iowa when he got into a dispute with a clergyman about the literal translation of the Bible. The clergyman insisted that such passages as in Genesis 6:4, which read, "There were giants in the earth in those days" was absolute fact. Hull strongly disagreed, but the argument gave him an idea.

People were gullible. They believed in giants. People would spend their hard-earned money on stupid things. Gullible people would pay to see what they thought was a real giant, and George Hull would be happy to take their money.

There was gypsum at a nearby construction site, and Hull gave the workers a barrel of beer to cut out a big block of the stone for him. The massive block was transported to Chicago, where a sculptor carved a 10-foot giant. The figure was contorted, as if the man died in pain. To add more realism, the figure was repeatedly struck with a set of needles to simulate the appearance of pores, and then ink and acid gave the surface an ancient, weathered appearance.

The next stage of the plan was to bury the giant on a relative's farm in Cardiff, NY. However, as many people had seen something big being transported along the route to the farm, Hull wisely decided to wait at least a year to make the "discovery." Fate was apparently on Hull's side, as about six months later, legitimate fossilized bones were found on a nearby farm. The stage was now set for the great Cardiff Giant Hoax.

William C. "Stub" Newell, on whose farm the sculpture was buried, hired some neighbors to help him dig a well near his barn. When their shovels struck something hard and they started to push away the dirt, one of the men at first thought they had found an old Indian burial site. After a little more digging, he realized it was unlike any Indian he had ever seen! While Stub pretended to be shocked, the other men were genuinely astonished.

Word of the discovery spread quickly, and right away a tent was erected over the giant man and a twenty-five cent admission was charged to the throngs of visitors who began arriving. To handle all the tourists, a stagecoach line set up four trips a day from Syracuse to Newell's farm. The admission fee to see the giant ancient man was quickly doubled and it looked like Hull's $2,600 investment was going to pay off. Then the scientists and experts began arriving.

Hull was sure the jig was up, but to his amazement they didn't say his giant was a hoax. Half claimed it was indeed a petrified man, while the other half declared it was an ancient stone sculpture—in either case, well worth the long trip and fifty cents admission.

The real payoff came when a group of investors, led by banker David Hannum, paid Hull over $30,000 to put the Cardiff Giant on display in Syracuse. As $30,000 in 1869 was roughly the equivalent of three-quarters of a million dollars today, Hull instantly became a wealthy man.

Enter Othniel Charles Marsh, a vertebrate paleontologist from Yale. Upon close examination of the alleged giant, Marsh found telltale chisel marks and areas where the ink and acid hadn't been sufficiently applied to create that ancient look. Marsh announced the Cardiff Giant was clearly a recent fake, a hoax, a crude fabrication made to fool the public and take money out of the pockets of the gullible. And how did people react to this news that they had been duped? They kept coming and paying money to see "Old Hoaxey" the fake giant!

(Clearly there was nothing else better to do in rural upstate New York in 1869. Come to think of it, there still isn't…)

The Cardiff Giant being raised from his "grave."

With all this money passing hands to see an obvious hoax, it was bound to attract the king of blatant fraud and deception, P.T.

Barnum. He wanted the Cardiff Giant for himself, but when his offer of over $50,000 was turned down, he had only one option—make his own. He hired someone to surreptitiously make a plaster replica, which he promptly claimed was the "real" fake Cardiff Giant—not that "fake" fake one in Syracuse—and he put it on display in New York City. Even more people paid good money to see this reproduction fake.

Upon hearing that Barnum's phony giant was making more money than his original fake, David Hannum uttered the immortal words, that, "A sucker was born every minute." That phrase was later attributed to Barnum, a mistake he probably neglected to correct, but then veracity was never his strong point.

The other thing Hannum did was to file a lawsuit against Barnum for the unauthorized copying of a fake, as well as Barnum's slanderous claim that Hannum's original fake was the reproduction fake. (This is getting confusing, isn't it?)

So, what happened to the principle players in this story? George Hull lost his fortune to bad investments. (Can you say bad karma?) He tried to "discover" another giant in Colorado, but this time the public didn't buy it. He eventually went to England and faded into obscurity.

Hannum's lawsuit was dismissed, as making a copy of an admitted fake wasn't a crime. Barnum continued to make boatloads of money with his Cardiff Giant and many other fakes and oddities. His version of the giant is now on display at Marvin's Marvelous Mechanical Museum near Detroit. The original fake is in the Farmer's Museum in Cooperstown, NY.

So, almost a century and a half after the enormous, bogus Cardiff Giant emerged from the dirt of an upstate New York farm, people are still paying to see both him and his twin!

Behold the enduring power of the hoax!

Another Giant Hoax

One would have thought that memories of the Cardiff Giant hoax would still be fresh in people's minds just ten years later, especially in a town only 50 miles away. But there truly is a sucker

born every minute (or second), and hotel owner John Thompson correctly surmised that what fooled people once could fool them again.

While Thompson gets an "F" for originality, he most certainly scores an "A" for composition and execution. Whereas the Cardiff Giant was carved out of rock, Thompson was far more ingenious, or at least the local mechanic he enlisted, Ira Dean, was, when he came up with a concoction of cow's blood, eggs, and iron filings all mixed together in some type of thick plaster. With this semi-organic material, Dean sculpted a 7-foot-tall "prehistoric" man, and then baked the massive figure until it hardened like a rock.

Obviously, just digging a hole and burying the giant man would elicit great suspicion when it was easily uncovered from the churned up earth. No, Thompson needed something cleverer if he was to draw more customers to his hotel in Taughannock Falls, New York. Workmen were widening a road on Thompson's property, so he and his associates (Dean and a third man, Frank Creque) carefully planned for these excavators to "discover" the prehistoric wonder.

Thompson and his two co-conspirators tunneled sideways to a spot they knew the workmen would be digging up. They managed to push the enormous 800-pound giant through the tunnel, and even wrapped some tree roots around it to give the impression that it had been there for ages. And lo and behold, on a hot summer day in July of 1879, one of the workmen was astounded when he uncovered the prehistoric petrified man.

As the Taughannock Giant was found on Thompson's property, it was his right to display it and sell photos to the thousands of curiosity-seekers who traveled far and wide to view this huge man from the distant past. In another stroke of genius, Thompson even allowed scientists from Cornell to chip off a few pieces of the giant for analysis, knowing full well that they would find substances consistent with animal blood and tissue. The scientists declared that the Taughannock Giant was indeed the real deal, which brought even more paying customers Thompson's way.

The lucrative scheme unraveled, however, when Creque had a little too much to drink. They say loose lips sink ships, and drunk ones sink hoaxes. Creque foolishly spilled the beans on the whole story. However, even with the truth revealed, the scientists who had examined and analyzed the giant still maintained that it was a real petrified man, and that Creque was lying! It wasn't until Dean mixed another batch of his prehistoric batter and baked a little giant for the scientists that they finally admitted they had been duped.

The fraud revealed, the flood of tourists dwindled to a trickle, then stopped altogether. As there's nothing more passé than a fake giant, it was decided to remove the figure from where it had been displayed. Unfortunately, moving an 800-pound hoax is easier said than done, and it was dropped and broke into pieces. The pieces were buried somewhere on the property, and the memory of the Taughannock Giant slowly faded.

Other hoaxers have subsequently created giants and other creatures, and even today good scientists are going bad by sticking bones of different dinosaurs together and claiming they have discovered new species. While modern technology has made it harder to pull off a successful hoax, there will always be those who will try to separate the gullible from their money…and more often than not, they will probably succeed.

Piltdown Man

One of the obstacles to firmly establishing the Theory of Evolution in the latter part of the 19th and early 20th centuries was the incomplete fossil record. The gap from the apelike creatures to the more human forms was bothersome, to say the least, and many sought the elusive "missing link."

Enter Piltdown Man, first formally introduced to the world in 1912. Far from being a complete skeleton, all that had been found was a section of an apelike jaw coupled with a large brain-capacity skull—indicative of a modern human—but it was more than enough to confound and amaze the scientific community. Many paleontologists hailed the find as the link that forged the unbroken chain of human evolution. The only problem was that it was about as real as the Pillsbury Doughboy.

From 1908 to 1912, Charles Dawson, who was a lawyer by profession, claimed that he was given pieces of the bones from someone working in the Piltdown quarry in England. However, there were those who believed that the skull and jaw—with a few worn, human-type teeth—were from different animals, and it was simply coincidence that they were found together. Healthy skepticism abounded until 1915, when more alleged discoveries surfaced.

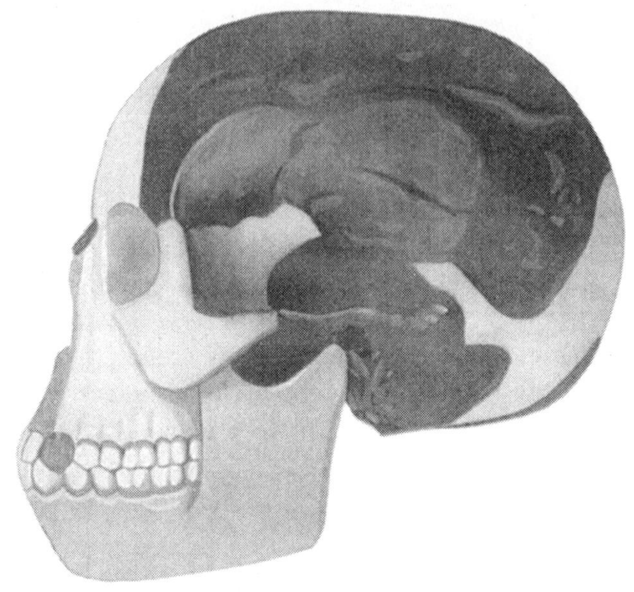

Sketch of the reconstructed skull.

Just when you thought it was safe to go back into the quarry, Piltdown II hit the scene. Skeptics who had discounted the first skull and jaw now also had to consider that the chances of two such sets of bones being found only two miles apart were beyond mere coincidence. Despite the lingering skepticism, however, some Englishmen proudly puffed out their chests and declared that the ancestor of mankind was British. Still, something just didn't look right…

The more bits and pieces of ancient hominids that were uncovered in Africa and Asia, the more Piltdown Man didn't seem to fit into the emerging scheme of things, yet still it held sway over some paleontologists. Then came a truly momentous and legitimate discovery, Australopithecus africanus, which had been painstakingly uncovered in South Africa by Raymond Dart. The skull fragments of the three-year-old Taung child, as it came to be known, were not quite ape, but not quite human, and dated to over one million years. Shaped animal bones found nearby led Dart to conclude that the child's "people" were toolmakers.

It was all remarkable new information, if true, and therein lay the dilemma. Those who doubted the authenticity of Piltdown Man cast an equally disbelieving eye at the Taung child, and those who believed in Piltdown Man claimed that it was Dart's discovery that was the odd link that didn't fit. So not only did the Piltdown Man waste precious research time and money, it also created obstacles to research into genuine fossils. While Dart was disgusted by the controversy, in the following decades new discoveries did establish Australopithecus as a vital part of the evolutionary chain, although the tool-making debate still continues.

The Piltdown problem persisted, however, for over forty years. While its importance gradually eroded until it was practically ignored, it wasn't until 1953 when the hoax was finally confirmed by Sir Wilfrid Le Gros Clark, Kenneth Oakley, and Joseph Weiner. The tip-off was the file marks on the teeth, which no one had previously noticed. The teeth had been reshaped to fit the jaw. Newly developed tests then showed that the jaw was actually only a 500-year-old orangutan bone. The skull was human, but found to be of an age of just 600 years—a far cry from an ancient fossil. Whoever perpetrated the hoax had treated the bones to make them appear to be ancient. And many trained observers fell for it all.

It was most likely Charles Dawson, himself, who produced the fake, and it now appears as though this was not his only hoax. Why his personal character wasn't examined more closely is a mystery (remember, he was a lawyer, after all), but perhaps it has something to do with British scientists being too eager to believe that the missing link had been found on British soil. Unfortunately, when the hoax was uncovered, Dawson was unavailable for comment—having died back in 1919—but he probably died laughing.

The Piltdown Man is now a universal symbol of fraud, gullibility, and the failure to be objective when something too good to be true comes your way. It's a shame that people feel compelled to trick their colleagues, as it is hard enough gathering evidence, conducting experiments, and drawing valid conclusions.

The writer Dante had the right idea. He placed souls who committed fraud in the Malebolge—the eighth level of hell just one flight up from Satan. Unfortunately today, those who fabricate and deceive are often placed in Congress…

"The phonograph has no commercial value at all."
Thomas Edison 1880s

"The radio craze will die out in time." Thomas Edison, 1922

"The wireless music box has no imaginable commercial value. Who would pay for a message sent to nobody in particular?"
David Sarnoff, American radio pioneer, 1921

"While theoretically and technically television may be feasible, commercially and financially I consider it an impossibility, a development of which we need waste little time dreaming."
Lee de Forest, inventor of the vacuum tube, 1926

"Television won't last. It's a flash in the pan."
Mary Somerville, radio broadcaster, 1948

"Television won't last because people will soon get tired of staring at a plywood box every night." Darryl Zanuck, movie producer, 1946

Missing the Link

Generations have learned about the world from *National Geographic* magazine. The National Geographic Society has funded important research and education for over a century, and has been a powerful voice for science. Unfortunately, even such venerable organizations as this can have a misstep, such as when they declared the discovery of a missing link in 1999.

One of the hotly debated issues in paleontology is the origin of birds, with two camps having formed. One supports the theory that they originated from dinosaurs, while the other camp maintains birds descended from a different group, perhaps a common ancestor of both birds and dinosaurs. The National Geographic Society was in the dinosaur camp, and when the perfect, 125-million-year-old missing link fossil appeared in Liaoning, China, they thought they had won their case.

The society funded the research conducted on the fossil of *Archaeoraptor liaoningensis,* "ancient bird of prey from Liaoning." The scientists chosen were Stephen Czerkas of the Dinosaur Museum in Utah and Xing Xu of the Institute of Vertebrate Paleontology and Paleoanthropology in Beijing. Their findings were announced in October of 1999 at a press conference in Washington D.C. At the same time, the November issue of *National Geographic* magazine featured the article *Feathers for T. rex? New birdlike fossils are missing links in dinosaur evolution.*

The press conference and article described this animal as having the straight tail of a dinosaur and the feathered body of a bird the size of a turkey. In glowing terms, the fossil was touted as "the true missing link in the complex chain that connects dinosaurs to birds. It seems to capture the paleontological 'moment' when dinosaurs were becoming birds," and its "mix of advanced and primitive features is exactly what scientists would expect to find in dinosaurs experimenting with flight."

(Uh, oh. If there's anything in science that should send up a red flag, it's finding *exactly* what you are looking for to prove your theory!)

The discovery that was to change "everything from lunch boxes to museum exhibits" was, in truth, quite a turkey, but one that would neither change lunch boxes or museum exhibits, and certainly not text books. *Archaeoraptor liaoningensis* actually captured the moment when hoaxers glued the fossil tail of a dinosaur to the fossil of a bird-like animal. Instead of discovering the missing link, the National Geographic team *missed the link* between the region where the alleged fossil was found and its notorious reputation for producing fakes.

In fact, people in the Liaoning province were well known for "enhancing" or just plain fabricating fossils by carving and cementing together fake creatures for fun and profit. It would be similar to going into an art museum's gift shop and buying a

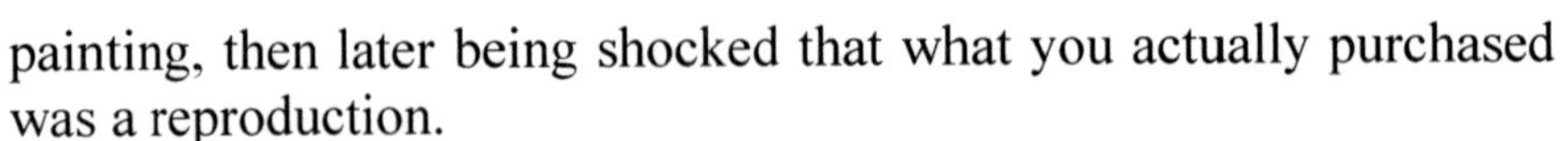

painting, then later being shocked that what you actually purchased was a reproduction.

Come on paleontologists, if you have just suddenly found *exactly* what you were looking for to prove your theory of the evolution of birds from a fossil "discovered" in a province with a reputation for fake fossils, shouldn't you be just a bit more cautious before making any announcements to the world?

In fairness, they did seek out other experts and they did receive several dissenting opinions. They just chose to disregard them.

Photos of the fossil had been sent to Storrs L. Olson, the curator of birds at the Smithsonian Institution. He disagreed that what the National Geographic team was calling feathers were actually the fossil remains of feathers. University of Kansas paleontologist Larry D. Martin, who specializes in bird fossils, couldn't find any feathers on the fossil, either. He even went so far as to point out that the fossil looked as though it had been pieced together from more than one animal, and that the all-important bones which should have linked the tail to the body were missing. Other scientists also noted the missing bones and evidence of fakery. Both publications *Nature* and *Science* rejected Czerkas and Xu's report.

Despite these additional red flags, the society proceeded with their announcement, which prompted Olson to declare in disgust that the society had done nothing less than "reached an all-time low for engaging in sensationalistic, unsubstantiated tabloid journalism."

A harsh, but not unfounded criticism for the century-old society, from which we expect the highest standards to be maintained.

The final nail in the coffin of *Archaeoraptor liaoningensis* came from Xing Xu himself. He went back to China in search of the other half of the fossil. (When a rock is split open to reveal a fossil, the two halves contain mirror images of the animal or plant.) It was no doubt a search that involved a lot of mixed feelings, and ultimately, disappointment and embarrassment ruled the day. Xu did find one of the fossil's counterparts, only this fossil dinosaur tail was still attached to its original dinosaur body. The bird/dinosaur missing link he had declared to be genuine had clearly been pieced together and was undoubtedly a hoax.

Criticism abounded, with the cleverest rebuke coming from *U.S. News & World Report,* who declared the *Archaeoraptor liaoningensis* to be the Piltdown Chicken. It seems that no matter what the era, you can fool some of the paleontologists all of the time.

Richard Stucky, vice president of the Society of Vertebrate Paleontology summed it up best when he said, "This is always an issue with commercially acquired specimens where economic values may supersede scientific concerns. Fabrications and enhancement are not uncommon and must always be considered when a specimen is acquired commercially rather than through a scientific expedition." (The fossil had been purchased for $80,000.)

We must also include that this is always an issue with commercially *and* scientifically acquired specimens, where egos, vanity, and the overwhelming desire to prove pet theories may supersede scientific concerns. Ultimately, however, the scientists duped by the Piltdown Chicken are not culprits, but victims—victims of forgers who for the sake of greed and some twisted sense of humor seek to continually sully the good name of science.

Such people do more than just ruffle a few dinosaur feathers with their fakes and fabrications, and we can only hope that they all become fossils very soon.

Divine Hands

Some people are born lucky, while it has been said that others make their own luck. Then there are those who make their own Stone Age artifacts and claim they are real—and for those people it's only a matter of time before their luck runs out.

Shinichi Fujimura was born in the Miyagi prefecture of Japan in 1950. As a boy, he found pieces of ancient Jomon pottery in his backyard, which sparked an interest in archaeology. While working in a factory, he began studying archaeology on his own and spent his vacations digging for artifacts. In 1975, he formed a group with other archaeology enthusiasts (Stone Age archeology is a very popular topic in Japan), and this group began discovering many Stone Age sites. Their first important find came in 1981, when Fujimura claimed to have found 40,000-year-old stoneware.

As this was the oldest stoneware ever found in Japan, the discovery worked wonders for Fujimura's career. But he didn't rest on his Stone Age laurels. Almost like magic, he and his team consistently uncovered older and older artifacts—and not just by a few measly thousands of years, but by hundreds of thousands. At over 180 sites, Fujimura's remarkable archaeological prowess earned him the reputation of having nothing less than "divine hands."

However, not everyone was drinking the Fujimura Kool-Aid. While few colleagues had the courage to speak against a veritable living legend, there were those scientists who publicly stated that Fujimura's "dubious claims" were based upon "flawed research." Apparently, enough suspicion was eventually cast that one of Japan's major newspapers, the *Mainichi Shimbun*, decided to investigate.

It was October of 2000, and the town of Tsukidate had become a tourist destination with international attention, thanks to Fujimura and his team's recent discovery of—you guessed it—some of the oldest artifacts ever discovered in Japan. This time the claim was that they had found indications of a dwelling that was 570,000-years-old! To commemorate the find (and make a lot of yen from tourists), Tsukidate made its official slogan "The town with the same skies viewed by early man." It even began selling a drink called—you guessed it—"Early Man." (I guess creating a drink called "Recent Fake" wouldn't have been as big a seller.)

In November, the Stone Age house of cards finally tumbled down when the newspaper published three shocking photos it had taken in secret. These photos showed Fujimura digging holes, placing artifacts in the holes, covering them with dirt, and later "discovering" them. Caught red-handed, Fujimura confessed during a press conference, claiming that he had done it because he had been "possessed by an uncontrollable urge."

Yes, it's called ego and greed.

The man with the divine hands admitted he had planted 27 of the 31 artifacts at a level that would make them appear to be much older. He also confessed to planting 61 out of 65 artifacts at another site, and 100% of the stonework from yet another site. However, as is common for charlatans who have been exposed, he still tried to grasp onto some shred of respectability by maintaining that those were the *only* times he faked results.

Sorry, Fujimura, your credibility ship not only sailed, it was torpedoed and sank to the bottom of the Bad Science Sea.

There may be those readers who now say, okay, he was humiliated and exposed as a fraud, but he really didn't hurt anyone but himself.

Au contraire!

How about all the text books in Japan that now had to be changed because they contained decades of bogus finds? How about the dedicated researchers and real scientists who had spent years studying Fujimura's work? Imagine how stunned and angered his colleagues were, such as Hiroshi Kajiwara, a professor at Tohoku Fukushi University, who said, "My 20 years of research are ruined…Why on earth did he do such a stupid thing?"

That would be because of ego and greed.

Fujimura lost his position as the senior researcher at the Tohoku Paleolithic Institute, and its director resigned. Museums began pulling all of Fujimura's artifacts from their displays. Fujimura "wanted to be known as the person who excavated the oldest stoneware in Japan," and as a result he ruined the professional lives of countless people, and cast a shadow of doubt over other's legitimate finds.

However, all of this pales in comparison to another tragic result—one man actually took his own life over the repercussions of the scandal. Hungry for more evidence of forgeries by other researchers, in 2001, some articles in a Japanese magazine asserted that Mitsuo Kagawa, a professor at Beppu University, had also faked his Stone Age discoveries. The shame and suspicion drove Kagawa to commit suicide, but in his suicide note he maintained his innocence. The court ordered the magazine to apologize, as it didn't have sufficient evidence to make the accusation, but the apology was too late to help Kagawa or his family.

There is an old saying that people would rather be lucky than good, and indeed, luck often plays a hand in science, sports, the business world, and many other facets of life. Unfortunately, those who perpetrate one act of fraud successfully are often greedy for more and continue to push their luck with false claims of ever-

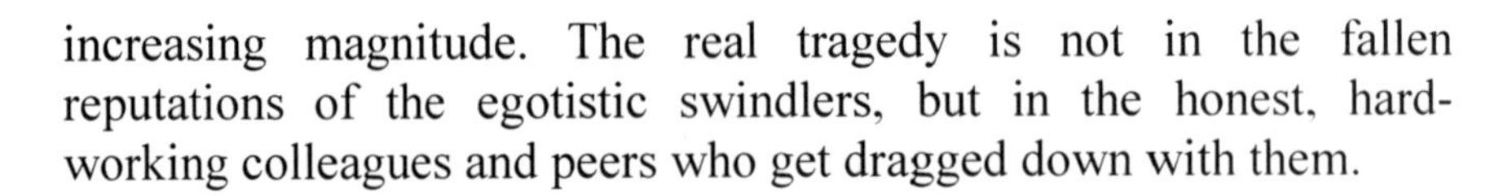

increasing magnitude. The real tragedy is not in the fallen reputations of the egotistic swindlers, but in the honest, hard-working colleagues and peers who get dragged down with them.

"[The automobile] will never, of course, come into as common a use as the bicycle." *Literary Digest*, 1899

The Thirst for Gold

For countless generations, gold fever has clouded the judgment of otherwise rational and prudent individuals. History has repeatedly shown that a person may be leading a completely ordinary life, but flash gold before their eyes and they suddenly drop everything and dash off on hazardous journeys to remote and dangerous locations to try to strike it rich in the latest gold rush.

It should come as no surprise then, that if in the late 1890s a humble store clerk, born and raised in a city, who never spent a day in his life riding or camping, was willing to risk everything by trekking to the Klondike in a search for gold, other people would be willing to risk their life savings to obtain gold much closer to home—without getting their hands dirty or even leaving the comfort of their homes. Such was the lure of the Electrolytic Marine Salts Company, which boasted it could extract gold from sea water and make every investor rich.

It all began in 1896, when a Baptist minister in Middletown, Connecticut, Rev. Prescott Ford Jernegan, claimed he had experienced a divine vision (that should have been the first clue that you were about to get hosed). In this vision, he saw how to make a device he called a Gold Accumulator. It was basically a wooden box with holes in it, and inside was a metal pan filled with mercury and some secret compounds, all of which had a current of electricity running through it. The idea was to submerge it in the ocean, let science do its thing, then in a few days or weeks simply retrieve the box and collect the gold that accumulated in the mercury.

Jernegan brought his device to the local jeweler, Arthur Ryan, and asked him to test it. If successful, i.e., if gold was produced,

Jernegan wanted Ryan to help him form a company, get investors, build facilities, and rake in the profits. While it was known that sea water contained traces of gold, there was no practical way to extract it in a manner that made financial sense. Ryan was skeptical, but agreed to test Jernegan's device. After all, would a clergyman lie about such a thing? (If you answered no to this question, please go back and start reading this book from the beginning and pay more attention this time.)

The most Christian Reverend Jernegan.

To prevent the appearance of anything underhanded, Jernegan said he would show Ryan how to set up the device and then he would leave. If the clergyman wasn't even there during the test, the device couldn't be tampered with, could it? (If you answered no to this question, please see above.)

Ryan and several associates took the device to a wharf in Rhode Island, where they lowered it into the ocean. Despite the cold February weather, the men remained on the wharf all night to guard the alleged Gold Accumulator and guarantee an honest experiment. When morning arrived, they raised the box and lo and behold, there were specks of gold in the mercury! When tested, it was found that the box had produced about $100 worth of gold by today's value. Certainly not a fortune, but what if a hundred, or a thousand, or tens of thousands of accumulators were at work?

Investors from New York to Boston didn't need their arms twisted to open their wallets, and the Electrolytic Marine Salts Company began operations in the remote Maine town of Lubec, ostensibly chosen for its vigorous tidal activity, which supposedly helped the accumulators accumulate.

The company's alluring prospectus stated that "Millions of dollars in gold were flowing through Lubec Narrows every single day," and their company had the means to extract it. The local newspaper, *The Lubec Herald*, praised Jernegan and his assistant, Charles Fisher, who said about the two men: "The presence of these people is not only desirable for the amount of money they will bring into the town, but we should welcome them for their social qualities. The officers of the company are earnest, Christian gentlemen, and many of their employees are Christians. We wish them all success in their undertaking and hope that they will take millions of dollars from old Passamaquoddy Bay – and we believe they will! With quantities of gold in the salt water there is little need of a trip to Alaska!"

A tide of people soon flooded the tiny town seeking jobs. Locals invested their life savings—even borrowing money and mortgaging their houses to be able to invest, because the news was good, and kept getting better.

In 1898, the newspaper wrote that "The inlet to Mill Pond accommodated 240 accumulators of which sixty were pulled up each week. Thus each box was under water a month before its turn came to be examined. During that time the water, chemicals, and electricity had time to work their magic."

Gold was being "harvested" at the rate of several thousand dollars per day in current value, and stock was offered in the company to expand even further. The stock initially opened at $33 a share, but within days was trading for five times that amount. There was now $400,000 to put into the expansion and it looked like easy street for everyone.

At least until Jernegan and Fisher disappeared in July of 1898, and the gold accumulators suddenly stopped accumulating.

Was foul play involved? Yes, but the missing men were most assuredly not the victims.

When Ryan and his associates spent that cold night on the wharf in Rhode Island making sure no one tampered with the box, they did not take into consideration what was going on under the surface. It seems that Fisher was an accomplished diver, and while Ryan and the other suckers were freezing their butts off on the wharf, Fisher was under water substituting the plain mercury for the gold-studded mercury.

Fisher also did some night diving in Lubec to add some gold to the accumulators that were scheduled for collection. To put icing on the phony cake, Fisher had also purchased bars of gold, which he then showed to the investors, passing them off as ocean-fresh specimens of the latest gold to be taken from the sea. Unfortunately, the only thing taken was the investors.

An article in the August 1, 1898 edition of *The Boston Herald* summed it up nicely: "Never did a fisherman bait his traps with more alluring or attractive morsels than did Reverend P. F. Jernegan tickle the fancy and stimulate the greed of victims with his brilliant and enticing prospectus of the Electrolytic Marine Salts Company, now a practically defunct organization, with the reverend promoter flown to foreign parts, an alleged swindle of the first magnitude."

Over 700 workers were suddenly unemployed, the life savings of hundreds of people had gone out with the tide, and the little town of Lubec, which had high hopes of becoming world famous for its vast wealth, faded into obscurity. (There are unconfirmed reports that today, even residents of the town never heard of Lubec.)

So, as lives lay in ruin throughout the northeast, where were the two swindlers? Fisher took his $200,000 and disappeared off the face of the earth. The most reverend Prescott Jernegan took his family to Europe—under an assumed name—with his cut of the money.

Apparently, however, the good Christian gentleman had a pang of guilt while living it up in Brussels. He decided to return money to the investors, but it must have been just a brief pang as he returned only $75,000. Some used this act to claim that Fisher was the mastermind behind the scheme and that Jernegan was a reluctant participant.

Returning all $200,000 is an act of remorse by a reluctant participant. Stealing and keeping $125,000 is not.

Jernegan eventually ended up in the Philippines, where he became a schoolteacher and the author of a book on the islands' history. He was never brought to justice.

As for all the investors, given the money Jernegan returned, plus what was realized in the sale of the company's assets, a mere 36 cents on the dollar was recovered. (Actually, only a 64% loss on an investment wouldn't be half bad in today's market, and surely good enough to earn those on Wall Street a hefty bonus!)

In all fairness to those who were duped, with all the marvels of technology the modern age has witnessed, it wasn't all that farfetched that a scientific method had been found to cheaply extract gold from the oceans. With gold fever raging from the Klondike gold rush, the time was right for just such a scam as Jernegan and Fisher's to succeed. Unfortunately, the moral of the story is that people will always be blinded by greed, and those who are ignorant of science are susceptible to the schemes of those who claim to wield its secrets.

While having reserves of gold isn't really a bad thing, ultimately, the best thing to accumulate is knowledge.

Catch My Drift?

When I was in the fourth grade, I distinctly remember a lesson in school that was fun (which is probably why it stands out from the usual mind-numbingly boring classes). The subject was continental drift, and each student was given a copy of a world map and a pair of those ubiquitous round-tipped scissors. The idea was to cut out the continents and fit them together as best we could into Pangaea—the original single landmass from which all the continents split and went their separate ways.

What had become mere child's play in elementary schools across the country was once a hotly debated—and resoundingly criticized—theory in the early 1900s. While as a nine-year-old, I was unaware of whom I had to thank for this exercise in cutting and fitting continental puzzle pieces. Now being older and wiser (as well as being in the possession of several *very* pointy pairs of scissors), I know that it was the German-born scientist Alfred Wegener.

Wegener earned his Ph.D. in astronomy in 1904, but almost immediately turned his attention to meteorology. However, that didn't satisfy his interest in geophysics, so he researched that topic on his own while he taught and wrote books on meteorology, before, during, and after fighting in World War I, where he was wounded twice. (Don't you just hate underachievers?)

Wegener noticed two major things that led him to his theory of continental drift: the coasts of South America and Africa looked as though they once fit together, and similar fossils were found in similar strata on opposite sides of the Atlantic Ocean, strongly suggesting that they were once part of the same landmass. The conventional wisdom (and I use that term loosely) was that ancient land bridges were responsible for the fossils—apparently very, very long land bridges that stretched across the ocean. Wegener didn't buy it, and in 1912 he first published his theory that the continents on either side of the Atlantic Ocean had once been together and were now steadily moving apart.

An illustration of an early landmass configuration.
(Not quite as good as my fourth grade Pangaea.)

He didn't know what the mechanism behind such massive movements could be, but undaunted, he continued his research and in 1915—not letting a little thing like the world war get in the way—he published the book *Die Entstehung der Kontinente und Ozeane*, otherwise known as *The Origin of Continents and Oceans*. The more he learned, the more he shared in subsequent revised and expanded editions of the book.

However, only one edition appeared in the United States in the year 1924. The book and its crazy ideas of moving continents was so harshly criticized by American scientists that it was decided not to bother printing any more editions. In fact, the president of the American Philosophical Society called it nothing less than, "Utter, damned rot!" (But how did he *really* feel?)

Alfred Wegener, c.1925

Another scientist in the U.S. bitterly complained that, "If we are to believe [this] hypothesis, we must forget everything we have learned in the last 70 years and start all over again."

(Yeah…so…isn't that what we call research and knowledge, which is the basis of science?)

But lest you think just those on the American side of Pangaea were the only ones who didn't catch the drift, a British geologist declared that any individual who "valued his reputation for scientific sanity" would never entertain such a ridiculous theory. The criticisms continued to rain down upon Alfred Wegener from all sides—including his own father-in-law, who happened to be Germany's leading meteorologist.

There were some scientists who did recognize the validity of his evidence and supported his theory, but those scientists were few and far between. In a somewhat lame defense of Wegener's critics, modern-day scientists have pointed out that he got the rate of drift wrong (Wegener's theories had the continents fairly cruising rather than drifting, with rates at anywhere from ten to one hundred times faster than is now believed), but an error in predicting how fast something moved should not discount the fact that *it did move*.

Unfortunately, Alfred Wegener did not live long enough to see his theories vindicated. He had a lifelong fascination with Greenland, and during his fourth expedition there in 1930 he was caught in a blizzard and froze to death. Almost fifty years after first presenting the ideas of continental drift and Pangaea to the world, the mechanism of sea-floor spreading was finally discovered—the mechanism that drove the landmasses and unlocked the origins of the continents and oceans.

Today, Wegener is not exactly a household name, but he has posthumously been honored by having craters on both Mars and the Moon, as well as an asteroid, named after him. And thanks to this courageous scientist, how many other schoolchildren, feverishly working their round-tipped scissors, will cut and paste the continents together, and so easily learn a lesson that the most educated men once called, "Utter, damned rot!"

Be Careful What You Ask For

In Japan on November 11, 1971, researchers from the Agency of Science and Technology conducted an experiment to study landslides. They soaked the side of a 60-foot hill using fire hoses, but failed to anticipate the 9-foot-high wall of mud and rocks that swept down the hillside, engulfing scientists and journalists alike. Fifteen people were killed, and nine were injured.

Hopefully, the surviving scientists realized that landslides are best studied at a distance.

The Miracle Mineral

The word asbestos is from the ancient Greek meaning "not extinguishable," due to its flame-resistant properties. Unfortunately, it does extinguish something—life.

It was the geographer and historian Strabo (c.64 B.C.-23 A.D.) who first wrote about the respiratory problems asbestos miners suffered on island of Euboea. Slaves who wore clothing made with fabric woven with asbestos also showed signs of lung disease. However, sick and dying miners and slaves weren't sufficient to stop the mining and use of asbestos, and unfortunately, thousands of years later, more lives would also be sacrificed on the altar of what came to be called the "Miracle Mineral."

Microscopic image of the deadly fibers of asbestos.

In the 1890s, asbestos was being mined around the world, and would eventually be used in a wide variety of products, including house shingles, tiles, insulation, appliances, car brakes, yarn, mittens, cigarette filters, and even baby powder! It takes about ten years for symptoms of asbestos-induced lung disease to manifest, and right on schedule the first diagnosed case of asbestosis occurred in London in the early 1900s.

The victim was a 33-year-old textile factory worker, and during his postmortem, the doctor found asbestos fibers in the man's lungs. Once these tiny fibers enter the lungs they become permanently embedded, causing irritation, inflammation, and scarring. As scarred lung tissue loses its elasticity, breathing becomes increasingly difficult, and eventually leads to death.

Another disease brought about by inhaled asbestos fibers is cancer, both in the form of lung cancer and malignant mesothelioma, a rare and particularly deadly form of cancer. Victims may not develop mesothelioma until decades after exposure to asbestos, but the link between asbestos and cancer was already evident in the 1920s. By that time, health insurance companies were denying coverage to people who worked with asbestos due to the "assumed health-injurious conditions" encountered in the factories.

Despite the mounting death toll and obvious link between asbestos exposure, respiratory disease, and cancer, over the ensuing decades the production of asbestos products grew in leaps and bounds, due to that age-old motivation—profit. Unfortunately, the value of human life didn't appear on the balance sheets of most big corporations.

Those same corporations also made sure other things didn't appear, such as medical evidence clearly illustrating the deadly hazards of the Miracle Mineral. Reports and studies were effectively blocked from being released to the public, such as the U.S. Bureau of Mines and Occupational Health Clinic study in 1932, conducted on workers at the Johns Manville factory in Oklahoma. The examinations included x-rays of the lungs of the employees and it was found that 29 percent had asbestosis. Johns Manville made sure the report was never released.

> Asbestos was used extensively in naval shipyards, during both World War II and the Korean War, to insulate pipes and boilers. Exposure was not only high for the men who built the ships, but for the crews stationed aboard them, as well as the longshoremen who handled the bundles of asbestos.
>
> Mount Sinai School of Medicine conducted a study that revealed that for those shipyard employees with at least 20 years on the job, **86 percent** of them developed lung disease or cancer due to their exposure to asbestos.

Years of bribery, threats, and lies continued to keep the truth from the general public, resulting in such suffering and death in a scope and scale that is unimaginable. But did corporate big shots lose sleep over their deadly deception? Hardly, as a memo from the Bendix Corporation to Johns Manville in 1966 illustrates: "…if you have enjoyed a good life while working with asbestos products why not die from it."

In 1952, the medical director of Johns Manville, Dr. Kenneth Smith, requested that warning labels be applied to products containing asbestos. His request was denied, and Dr. Smith later testified as to the company's reasoning: "It was a business decision as far as I could understand…the corporation is in business to provide jobs for people and make money for stockholders and they had to take into consideration the effects of everything they did and if the application of a caution label identifying a product as hazardous would cut into sales, there would be serious financial implications."

> "We feel that the recent unfavorable publicity over the use of asbestos fibers in many different kinds of industries has been a gross exaggeration of the problems. There is no data available to either prove or disprove the dangers of working closely with asbestos."
>
> A Raybestos-Manhattan company official, 1966

Countless other evidence of asbestos companys' blatant and callous disregard for their workers' safety has been revealed in the last few decades as many asbestos exposure-related lawsuits were filed. For example, in 1984, Charles Roemer, a manager at Johns Manville in 1943, testified that in that year, the president of the company stated that a competitor's managers were "a bunch of fools for notifying employees who had asbestosis." Roemer then said to the president, "Do you mean to tell me you would let them work until they dropped dead?" The president's response? "Yes. We save a lot of money that way."

The truth finally emerged and the corporations who knowingly exposed their workers, and then suppressed the deadly evidence, have at least been paying billions in financial settlements for their crimes—although despite what big corporations think, money is no substitute for health and life.

In 1979, the Environmental Protection Agency announced it would be banning almost all uses of asbestos, but it was not until 1989 that the ban took effect. Remarkably, even after all that has happened, asbestos companies won a federal lawsuit in 1991 that repealed that ban! Although greatly limited, the use of asbestos continues to this day, to the tune of millions of pounds being used every year.

Though shocking in their human toll, these asbestos horror stories should not surprise us, and we should never assume this was an isolated example. How many other cover-ups for profit are taking place around the world at this very minute? How many lives will be ruined by toxic substances that right now are being ingested, breathed in, or entering our systems through contact with plastics, fabrics, cosmetics, and hundreds of other products? What damage is now being caused by the electromagnetic fields of cell phones and dozens of other handheld and desktop devices with which we surround ourselves on a daily basis?

As long as people with no conscience value profit over safety, this asbestos story will play out again and again in countless variations, over countless generations. If people believe corporations and the government will protect us, then we are all indeed "a bunch of fools."

Monosodium Glutamate (MSG) is widely used as a flavor enhancer, yet in the 1950s it was already proven to damage the retina, and later shown to also kill neurons in the brain. MSG is recognized to be an excitotoxin—a group of substances that destroy nerve cells through excessive stimulation. Numerous food additives have excitotoxic effects.

Aspartame, marketed as Nutrasweet and Equal, may cause over 100 side-effects, from migraines, to loss of vision, to psychological problems such as anxiety, depression, and insomnia. Yet, aspartame is used in over 5,000 products such as diet soda and gum, and is consumed by over 250 million people.

Children are particularly sensitive to the damaging effects of food additives.

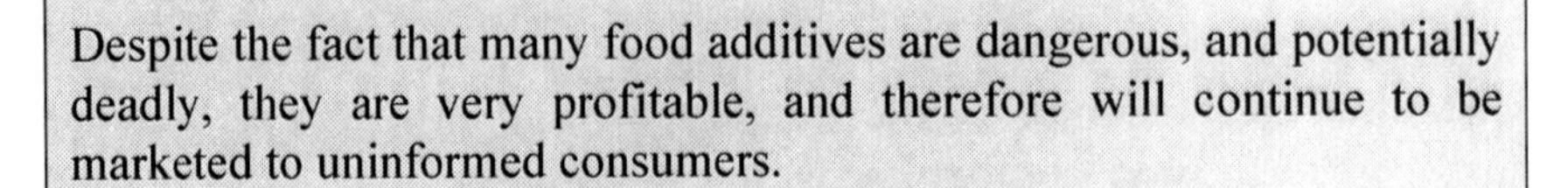

Despite the fact that many food additives are dangerous, and potentially deadly, they are very profitable, and therefore will continue to be marketed to uninformed consumers.

The King of the Hoax

P.T. Barnum’s ad for the lovely Fiji Mermaid, and the actual hideous “mermaid” that was a combination of a fish, a baby monkey, and some papier-mâché—the perfect example of great marketing triumphing over science for fun and profit.

FEJEE MERMAID,

which was exhibited in most of the principal cities of America, in the years 1840, '41, and '42, to the wonder and astonishment of thousands of naturalists and other scientific persons, whose previous doubts of the existence of such an astonishing creation were entirely removed ;

Astronomy & the Space Program

Just in Time for Dinner

For thousands of years, mankind has pondered the age of our world, but few individuals felt themselves clever enough to try to pin Creation down to the exact day.

Enter Archbishop James Ussher, Primate of Ireland (a fitting title), born in 1581. Considered a scholar in his time, Archbishop Ussher took the Bible and tallied up all the generations listed back to Adam and Eve, basically compiling a "Who Begot Who" of the ancient world. Then, employing what were undoubtedly painstaking calculations, he determined that the world came into being in the year 4004 BC.

Not content to simply bull's-eye the year with startling accuracy, the sage Archbishop further declared that Creation occurred on October 22, sometime in the evening—presumably giving Eve time to whip up a little dinner before she and Adam tottered off to bed in their new home.

Remarkably, his calculations were widely accepted—even Sir Isaac Newton believed the Creation date of 4004 BC to be accurate to within an error of only twenty years. In fact, this ridiculous date did not meet any serious challenges until the late 18th century, when works of geologists like James Hutton finally established that our Earth, and therefore the universe, was far older than the primates dared imagine.

"We are probably nearing the limit of all we can know about astronomy." Simon Newcomb, astronomer, 1888

A Slight Miscalculation

To be an epicurean is to seek pleasure, not in the overtly sensual hedonistic sense, but in a refined, genteel manner. This lifestyle got its name from Epicurus, a Greek philosopher who was born on the island of Samos in 341 BC. Believing that mankind was unnecessarily troubled by the fear of death and retribution from the gods, Epicurus tried to ease men's minds by explaining that their fears were groundless. He also developed a detailed outline for living life in a manner that reduced pain and increased pleasure, without harming anyone else. In a world half a step away from barbarism, striving for happiness wasn't such a bad way to pass the time.

However, Epicurus should have stuck with his happy philosophies, for when it came to astronomical theories, it turned out to be a bad way for him to pass the time. For example, despite the fact that some scholars of the era correctly maintained that the world was round, Epicurus believed that it was flat. But if an individual really wanted to believe it was round, that was okay with him, too, because you should believe whatever makes you happy.

Another of Epicurus' beliefs should put a smile on everyone's face. When asked how large was the sun, he declared it was no more

than two feet in diameter. Yes, that was no more than *two feet* in diameter. Now remember, Epicurus wouldn't be happy if you laughed so hard it hurt.

Not So Happy Hour

The age of Pericles in Athens during the fifth century BC is generally regarded as the pinnacle of Greek civilization. Adept in all areas from the martial to fine arts, Pericles was the symbol of courage, wisdom, and the democratic spirit. Surprisingly, however, this time of prosperity, power, and learning also paradoxically saw the banning of the study of astronomy. But it will come as no surprise that this act was religiously (i.e., politically) motivated.

Such were also the motives behind the condemnation of Pericles' friend, the philosopher and scientist Anaxagoras. Arriving in Athens at the age of twenty, the young Anaxagoras shunned the pursuit of wealth in favor of the pursuit of science, going so far as to state that the purpose of life was to study the Moon and stars.

During his lifetime, he made some remarkable (and correct!) assertions: Eclipses of the Moon were caused by the shadow of the Earth, the surface of the Moon contained mountains and deep valleys, and its light was reflected from the sun. He always strove to find a natural explanation for everything and even concluded that because man stood erect and could use his hands to examine things, our species developed superior intelligence.

Unfortunately, this evolutionary theory of intelligence did not seem to extend to the leaders of Athens, who indicted Anaxagoras for his blasphemous theory of the sun. The simplest mind in Athens knew that the sun was obviously a god, yet Anaxagoras had rashly used his inquiring mind to hypothesize that the sun was actually fiery, superheated rock, which, he boldly added, was probably many times larger than the entire Greek peninsula of Peloponnesus. The political enemies of Pericles seized the opportunity of eliminating their rival's friend, arrested Anaxagoras, convicted him of atheism, and threw the aged scholar in prison.

Fortunately, Anaxagoras was able to avoid the Athenian cocktail party Socrates would later be compelled to attend. While accounts vary, it seems that just at the appointed hour of execution, Pericles aided Anaxagoras' narrow escape, even as the hemlock was being shaken, not stirred.

Counter-Intelligence

The great Pythagoras (582-507 BC) left no writings of his own, and it was almost one hundred years after his death before a follower of his teachings, Philolaus, began the task of putting down Pythagoras' ideas for posterity. While he was at it, Philolaus decided to put down a few theories of his own.

The following summary will in turn, put down those theories... as far down as possible.

While Pythagoras had taught that the Earth was a sphere, he erroneously placed it firmly in the center of the universe and declared that everything revolved around our world. Philolaus didn't like the idea of the Earth at the center, not for any evidence based upon observation or calculation, but simply because he thought the Earth was too unrefined to inhabit such a noble position.

If our planet was not in the center of the universe, then Philolaus' theory necessitated the monumental step of setting the Earth in motion. This he did, but not around the sun. Being the motivated planet-starter that he was, Philolaus created some new heavenly objects to fit his scheme of things. At the center of everything he placed the "central fire" (also called the "watchtower of Zeus" or the "hearth of the universe"). Around this central fire revolved the Earth, sun, and all the planets and stars.

Of course, no one had actually ever seen this fire, so Philolaus needed to explain why. This time his solution was even more ingenious—in order to explain why we can't see the central fire, he created another object we can't see. This second object was a new body he called the Counter-Earth, or Antichthon, and it conveniently solved two problems.

First, even though the fire was never visible from Greece, because the Greek side of the Earth never faced the center of our orbit (where the central fire was allegedly located), people on the other side of Earth couldn't see it either, because their view was blocked by the Counter-Earth. Thus, Philolaus resorted to an expedient, yet self-defeating, theory which at the time was impossible to either confirm or deny. (Perhaps he should have abandoned astronomy for politics?)

The second problem the Counter-Earth seemed to solve was that of pure numbers, something with which the Pythagoreans were deeply enamored. Pythagoras felt that there was a harmony in the universe that could be expressed through mathematics, and numbers took on an almost mystical quality. Ten was a very important number to the Pythagoreans (apparently because it was the sum of the first four numbers, which personally doesn't send a chill up my spine, but hey, I'm not Greek) and Philolaus was thrilled that his addition of the Counter-Earth brought the total number of heavenly bodies to ten (the sphere of stars was considered one of the bodies).

There is never any mention about the Counter-Earth being inhabited, but it is likely that Philolaus assumed it would not be, because only the region between the Earth and the Moon was thought to be capable of sustaining mortal beings. However, he did believe the Moon to be inhabited with flora and fauna similar to that of Earth's—except for the "fact" that the Moon's creatures were fifteen times stronger, ostensibly due to their day being fifteen times longer.

Fortunately, Philolaus' theories did not survive for very long after him, but for a time there were some who followed his lead and created still more bodies we can't see. For some odd reason, there is a pattern throughout astronomy, as well as other sciences, that before an erroneous concept can be expelled it has to be expanded upon ad nauseam, employing the ever popular technique of counter-intelligence.

The More The Merrier

Most people can recall a time at school when they did something to aggravate a teacher. If that teacher happened to be Plato, the incident would be exceptionally memorable.

Eudoxos of Cnidus studied at Plato's famous Academy and on one occasion irritated the great teacher by conducting an experiment on mechanics. Plato was annoyed because experiments "corrupted" the purity of philosophy. Why soil a beautiful scientific concept with dirty experimental data? With this mindset drilled into every pupil, the scientific community of the 4th century BC too often took one step forward and two steps back.

Eudoxos eventually went on to form his own school and became famous for his knowledge of many subjects. One of his theories in particular, however, really helped put Cnidus on the map, and it involved astronomy. The prevailing belief in educated circles was that the heavenly bodies were attached to solid, crystalline spheres, which, being made of clear crystal, conveniently rendered them invisible to our eyes. (Where have we heard that before?)

However, from what was known of planetary motion, the planet/sphere concept clearly failed to explain actual observed motion (what little they bothered to do). This troubled Eudoxos, but rather than gather more data to form an alternate theory to the spheres, he added more spheres.

A lot more.

The "Spheres of Eudoxos" was a revolutionary (and incredibly complex and completely ridiculous) theory consisting of 27 nesting, rotating spheres. The planets each had 4 spheres, but the Moon required only 3. For example, the system of spheres for the Moon worked as follows:

- The innermost sphere rotated in a west to east direction, taking 1 month to complete a rotation.
- The middle sphere rotated on a different axis and in the opposite direction, moving from east to west with a period of 18 years.
- The outermost sphere also rotated from east to west and did so once a day.

This theory was hailed as being ingenious and elegant—so elegant, in fact, that no one seemed to mind that it completely failed to explain eclipses, the orbit of Mars, and many other observable phenomena, had they felt it important enough to observe.

As centuries passed and the inaccuracy of the 27 spheres became increasingly apparent, it was clear that a new theory was required. The great Ptolemy in the 2nd century A.D. turned his talents to the

 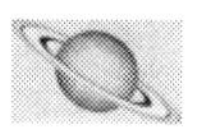

problem and came up with a remarkable remedy—he added more spheres!

A lot more.

By the time Ptolemy was finished, he had "fine-tuned" the system up to 80 spheres. The thinking must have been that if 27 was elegant, 80 was nothing less than divine. While there were to be several more variations on the theme, the basic Eudoxos/Ptolemy—type system persisted for another 1500 years, until the novel concept of creating theories based upon observational data finally shattered the crystalline spheres.

> I much prefer the sharpest criticism of a single intelligent man to the thoughtless approval of the masses. Johannes Kepler

Job Security

For over two thousand years, Western civilizations viewed the movements of the universe in terms of solid, crystalline spheres. There were many different theories regarding the number of spheres involved, and equally as many explanations for the *primum mobile*, or force which kept the spheres moving.

Some of the ancient Greeks believed that the planets were moved by the various gods associated with the celestial bodies. Aristotle assigned the task to spirits. By the Middle Ages, most astronomers agreed that it had to be angels responsible for the awesome feat of keeping the heavens in motion for all eternity.

While some thought that an angel was necessary for each sphere or planet, a unique concept evolved which unfortunately eliminated a lot of angels' jobs. In an early Renaissance world enamored with gears and gadgets, it was envisioned that only one angel was necessary to move the spheres—an angel whose sole task it was to turn a crank that moved the gears that spun the heavens.

However, once Newton explained the laws of gravitation, even that one angel wasn't needed any longer.

It appears that no one's job is ever safe.

A fanciful depiction of the mechanisms of the universe.

Too Much of a Bad Thing

Few books have enjoyed the reputation and longevity of Ptolemy's *Almagest*, the text on astronomy that was used for over a thousand years. Even the name sings its praises. Originally titled *The Mathematical Composition*, the Arabs dubbed it "The Greatest" or "al megiste" from which it received its present name. The *Almagest* was one of the most sought after books in Europe as the continent emerged from its intellectual coma of the Dark Ages. To be sure, it contains a great wealth of knowledge. Unfortunately, it also contains many errors—errors that literally would take the blood, sweat, and tears of generations to correct.

When picking up the *Almagest* for the first time, it is easy to become overwhelmed by the many pages of equations, tables, and diagrams contained in its thirteen books. The sheer length of the work is also impressive, even though in the preface, Ptolemy states

that "in order not to make the treatise too long we shall only report what was rigorously proved by the ancients," yet continues by adding he will also attempt "perfecting as far as we can what was not fully proved or not proved as well as possible." With all the bases covered, Ptolemy launches into a series of statements regarding the nature of the universe, followed by his "proof" of their validity.

Even in Ptolemy's day (100—178 A.D.), there were those people who believed "the movement of the stars to be in a straight line to infinity." This idea Ptolemy promptly discounts and goes on to explain why the movement of the heavens must be spherical. Part of the explanation involves the belief in the ether—that most perfect of elements in which the lofty stars abide. Ptolemy argues that such a pure substance would naturally take on the perfect form of the sphere, hence the universe is a sphere and moves around in a circle.

With the concept of a solid, crystalline sphere of fixed stars encircling the etheric heavens firmly established in Ptolemy's mind, he continues by offering "proof" that the Earth is in the center of the universe. To him, this assumption is basically self-evident, for "all observed order of the increases and decreases of day and night would be thrown into utter confusion if the earth were not in the middle."

To round out the Ptolemaic view of the universe, it is essential that the Earth remains stationary and everything else in the heavens moves around us. To those who believed that our planet rotated on its axis or even moved through the heavens, Ptolemy counters that if this was the case, then "animals and other weights would be left hanging in the air, and the earth would very quickly fall out of the heavens." Obviously, he concludes, "Merely to conceive such things makes them appear ridiculous." (Yes, Ptolemy, something here is ridiculous, but not what you think.)

Some may argue that the errors of the Ptolemaic universe are excusable given the times and the lack of technology. However, such arguments don't hold up when considering people like Aristarchus, who proposed a heliocentric theory 500 years before the time of Ptolemy. It is also clear throughout the *Almagest* that not everyone agreed with his theories, and they actually held opposing beliefs that were often correct.

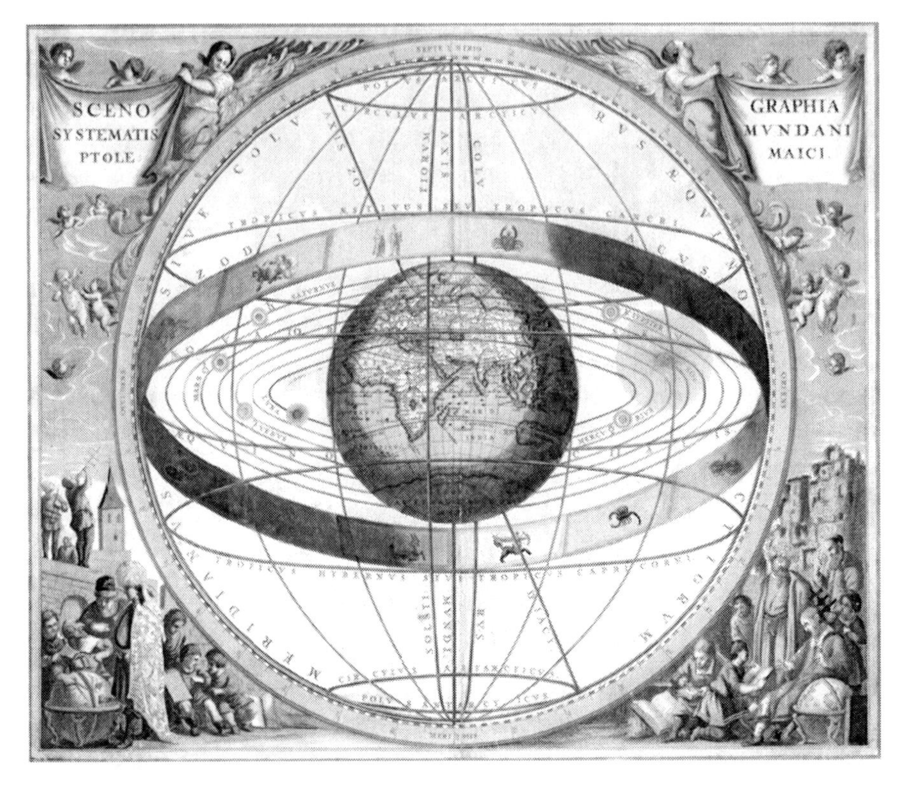

A 1660 depiction of an Earth-centered universe based upon the Ptolemaic System.

Yet history (which is usually just as blind as justice) favored Ptolemy and the *Almagest* with all its errors, and the work became *the* definitive text on astronomy until the time of Copernicus, when even then the transition to truth was long, difficult, and full of suffering for proponents of a heliocentric universe.

The story of that transition from Ptolemaic to Copernican systems really embodies the essence of Bad Astronomy. It brought out the best and worst of science, as well as religion, politics, and human nature. In removing the Earth from the center of the universe, mankind needed to redefine itself in terms in which it could thrive intellectually, creatively, and spiritually.

Perhaps the transition is even now not yet complete.

> To know that we know what we know, and to know that we do not know what we do not know, that is true knowledge. Copernicus

Music To His Spheres

In 1517, when Martin Luther nailed his ninety-five theses on the church door in Wittenberg, he was knocking on the door to a new age. While the Reformation needed the effort of many people to push open those hinges rusted with the ignorance of millennia, arguably the greatest single push came from an astronomer, Nicholas Copernicus. Presenting his heliocentric theories in *De revolutionibus orbium coelestium*, Copernicus lit the fuse to scientific dynamite.

However, while removing the Earth from the center of the universe made sound, scientific sense, many astronomers still clung to the old belief of the crystalline spheres and tried to reconcile them with Copernican theory. Ironically, the man who probably tried the hardest to do this would be the man who ultimately eliminated the spheres, Johannes Kepler. He flattened the theory, literally, into ellipses[1] and started a revolution of his own—but not without first flirting with Bad Astronomy.

Perhaps Kepler's deepest desire was to find a celestial harmony, regardless of who revolved around what. He supported Copernican theory, in part because it seemed to support such a harmony. According to what was known about the size of the planets' orbits, i.e., the size of the alleged spheres, Kepler believed that he had discovered that the five classic geometric shapes (tetrahedron, cube, octahedron, dodecahedron, and icosahedron) fit in between the five spheres. Instead of the ancient practice of adding more spheres to spheres, Kepler at least added a modern touch by inserting squares and triangles, and he published his findings in *Mysterium Cosmographicum* in 1596.

While this endeavor seems completely pointless to us today, the work was applauded by Galileo and Tycho, and it brought Kepler a fair amount of fame.

A strange paranoia seemed to pervade this age of astronomical discoveries—the amount of new scientific knowledge appeared to be directly proportional to the almost desperate attempts to find a divine blueprint, to assure mankind that the universe was not the

---

[1] Kepler discovered that the planets moved in elliptical orbits, not perfectly round orbits.

result of a chaotic, random happening. And if perfect spheres and other geometric shapes provided some level of comfort and a sense of serenity, then great minds would focus on this important task.

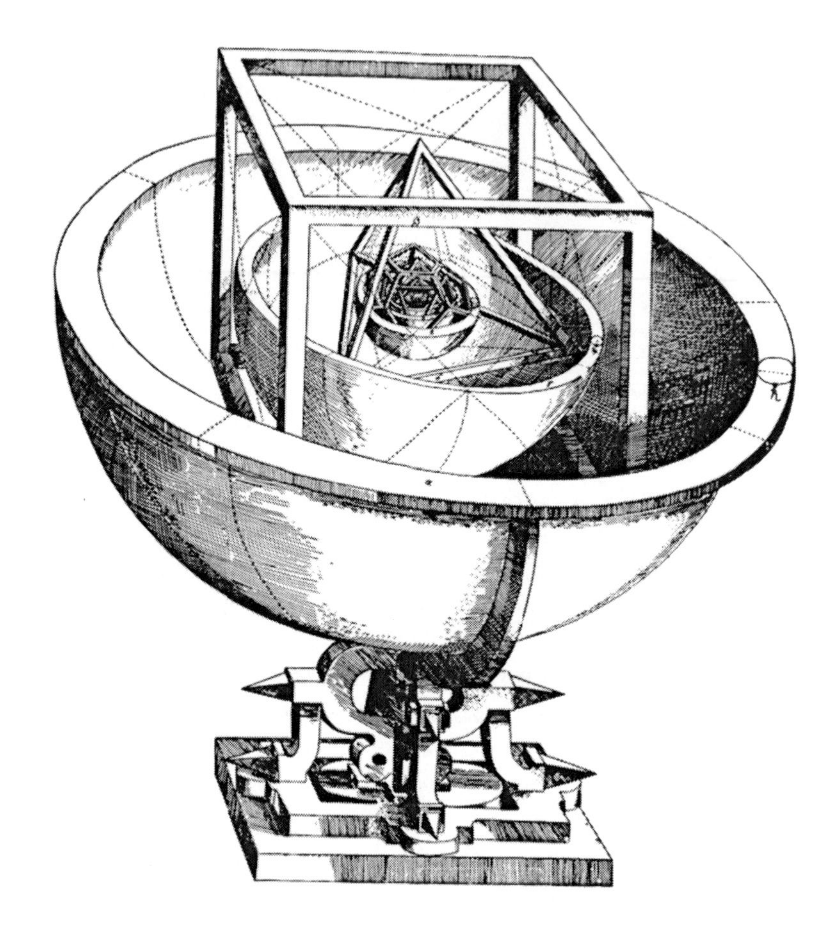

An illustration from *Mysterium Cosmographicum* of Kepler's attempt to insert geometric shapes into the solar system.

The harmonic bubble burst, however, when Kepler began studying the observational data collected by Tycho. To his credit, Kepler eventually abandoned his beloved celestial geometry because the facts dictated that he should. In his first two laws of planetary motion, he described the elliptical orbits of the planets and their changes in velocity, and provided the mathematical foundation for Copernican theory. Despite this great achievement, however, he still

wanted to find that elusive harmony. Once again, he thought he found it.

Kepler's third law of planetary motion reads: The squares of the periods of revolution are proportional to the cubes of their mean distances from the sun. To us this sounds like dry mathematics, to Kepler it sounded like divine music. After discovering this relationship, he transposed orbital velocities to a musical scale and he believed he had finally found God's blueprint. This was the actual "music of the spheres" Pythagoras had mentioned 2000 years earlier, and Kepler thought he had the equations to prove it. This apparent celestial harmony was what Kepler considered would be his greatest legacy to mankind.

Of course, history seldom runs according to plan, and few today have even heard of Kepler's great harmony, let alone believe it. Fortunately, Johannes Kepler will always be remembered for being the first man to accurately describe planetary motion. And for those who care to dig a little deeper, they will realize that what is Bad Astronomy to us, was music to Kepler's spheres.

Author's Note

The idea of celestial harmony or music of the spheres may sound like new age nonsense, but in fact, this search for harmony is probably as old as civilization, perhaps even as old as mankind itself. When it comes to trying to quantify this harmony in mathematical terms as Kepler did, I have to confess to being sympathetic to his cause.

Personally, I find mathematics very seductive. This is not to say that I turn the lights down low, slip into something more comfortable, and stare longingly at a series of equations. The attraction is primarily intellectual, but there is *something* (call it what you will) deeper.

It's difficult to describe in words, but if you've ever listened to a piece of music by Bach or Mozart you may have had some sense of their work being a beautiful melding of art and mathematics—a kind of auditory mathematics. Similarly, there is a visual mathematics to gazing at a Michelangelo sculpture or a painting by Raphael. I honestly believe that poetry, architecture, or any work of fine art is also a work of fine mathematics, and we are drawn to the unique harmony they represent. It isn't necessary to be consciously aware of the exact formulas and numbers involved—I think our brains can recognize structures intuitively.

So when a sensitive and brilliant mathematician like Kepler was studying a work of art as magnificent as the universe, how could he not feel that there was some all-pervading harmony throughout its structure?

It Takes an Upstart...

One might think that if the Church of Rome opposed the Copernican system in the 16th century, then Protestants would have accepted it. It is unfortunate, however, that this was one area where the two were in rare agreement. Theologians on both sides adhered to the literal meaning of passages in the Bible that referred to the Earth as being stationary.

Even Martin Luther, a man so eager to make changes with everything else, wanted to keep the Earth at rest in the center of the universe. In his own inimitable style, he referred to Copernicus as an "upstart astrologer." He further stated that, "The fool wishes to upset the whole science of astronomy, but the Holy Scripture shows, it was the sun and not the earth which Joshua ordered to stand still."

It just goes to show that it takes an upstart to know one.

Hitting Bottom

Kosmas Indicopleustes was born in Alexandria in the 6th century A.D., and after a career as a merchant, entered the service of the church. He had earned his last name (the meaning of which is "Indian navigator") as a result of his extensive travels. Having seen so much of this round Earth, Kosmas should have realized that it was not flat. Nonetheless, he steadfastly maintained that view, as well as maintaining a host of other unusual theories that also contradicted what he had seen with his own eyes.

Detailing his astronomical beliefs in the twelve-volume work *Christian Topography* (written between approximately 535-547 A.D.), Kosmas explained that the structure of the Earth and heavens

could be found in the Biblical description of Moses' tabernacle. For instance, the table within the tabernacle was rectangular and placed lengthwise from east to west. From this, Kosmas surmised that the Earth was a flat rectangle, oriented in the same direction. He also went on to describe the two-story framework which formed the regions of heaven, whose walls and curved ceiling he compared to that of a bathroom!

His primary goal throughout the twelve books seems to have been to disprove all Greek (i.e., pagan) theories, and on occasion after a thorough refutation, Kosmas appears to have been at somewhat of a loss for a better explanation. In the case of the crystalline spheres and the vehicles used to propel them, Kosmas blasts the Greek theories and warns that it is blasphemy for a Christian to entertain such absurd ideas. His scientific solution? Angels. Angels who until the end of time must carry the celestial bodies along their courses.

As for the movement of the sun and the explanation for the varying amounts of daylight throughout the year, Kosmas really outdid himself. His theories were based upon the well-known "fact" that everything was higher in the north than the south; the flat Earth was actually slanted. "Proof" of this "fact" was that ships took longer going north because they were essentially *sailing uphill*, and conversely, had a relatively easy and swift passage on the downhill or southerly course. (How many things are wrong with just that sentence!?)

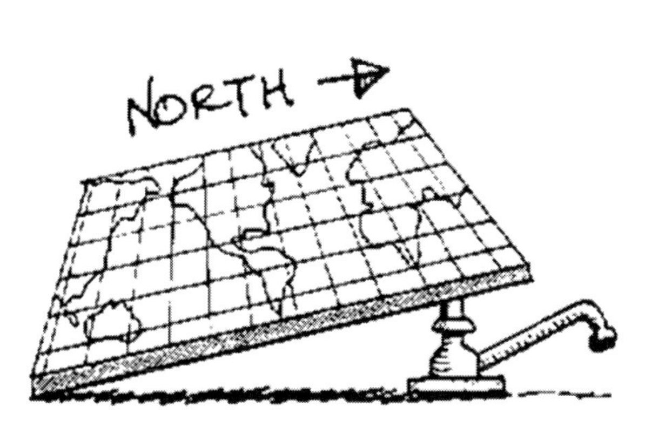

Joseph Gyscek

With this slanted view of the world, Kosmas took the ancient beliefs that the sun and Moon were returned to their positions in the east everyday by being carried along the north (the higher region, which blocked their light) and added his own twist. To account for the different lengths of summer and winter days, he theorized that a huge, pointed mountain high up in the north would allow more or less light depending upon at what point the sun was carried behind it. In essence, if the sun crossed near the narrower summit, the hours of daylight would be longer. Winter occurred when the sun was carried behind the broader base, thereby shortening the day.

The key to this theory was that in order for a mere mountain to block the sun, the sun had to be relatively small, so Kosmas conveniently calculated that the sun was actually only a little more than a thousand miles in diameter (significantly less than the actual 865,400 miles). And so in this manner, Kosmas continued through a dozen volumes, shooting down pagan theories and constructing his own politically correct replacements.

To his slight credit, on at least one occasion he did come up with a correct assumption, albeit for a terribly incorrect reason. The Earth, he declared, is most definitely not in the center of the universe. It could not be at the center, Kosmas asserted with confidence, because it is so heavy it must have sunk to the bottom.

A High Stake Gamble

Giordano Bruno was born in Italy in 1548. He joined the Dominican order, but because of a propensity for having his own opinions and daring to voice them, he was accused of heresy and left the order at the age of twenty-eight. For the next decade and a half, Bruno traveled across Europe learning, speaking, and publishing his thoughts. All the while, members of the Inquisition in Italy anxiously awaited the return of their hometown boy.

Among Bruno's outrageous beliefs were his assertions that the stars were actually other suns spanning the infinite reaches of space, and the entire universe was composed of the same matter. As if that wasn't bad enough, he firmly believed in the heliocentric theories of Copernicus, and said so in a voice so loud they could hear him in Rome.

And as if to prove how foolish he was, Bruno returned to Italy. Arrested, tried, and convicted, he was imprisoned in the dungeon of Castel Sant'Angelo in Rome for six years. Refusing to recant, he was finally treated to a barbecue, Inquisition-style, in the year 1600. While his heretical beliefs of astronomical theories were not the only crimes for which he was burned at the stake, they most certainly fueled the fire from which scientists such as Galileo no doubt smelled the smoke.

A defiant Giordano Bruno on trial before the Inquisition.
Bronze relief by Ettore Ferrari (1845-1929), Campo de' Fiori, Rome.

A Banner's Day

One of the most heinous crimes against humanity is for the powerful few to deny the subjugated many access to knowledge.

Submitted for your disapproval:

The *Index Librorum Prohibitorum*, or list of prohibited books—a handy little tool of the Inquisition, designed under the guise of piety to keep minds closed and collection boxes open and overflowing.

Two of the blackest dates for science were February 24, 1616, the day that Pope Paul V and the Inquisition decided that heliocentric theories were heretical, and March 5, the day Copernicus' *De revolutionibus* (printed 73 years earlier) was placed on the *Index*. (Actually, the book wasn't strictly forbidden. The

Church would allow it to be reprinted once it was "corrected," i.e., once the facts were sufficiently perverted. To all the printers' credit, no one chose to publish an altered version.)

INDEX LIBRORVM
PROHIBITORVM,
CVM REGVLIS CONFECTIS
per Patres a Tridentina Synodo delectos,
auctoritate Sanctiss. D.N. Pij IIII,
Pont. Max. comprobatus.

VENETIIS, M. D.LXIIII.

A 1564 version of the dreaded *Index*.

Copernicus' book wasn't the only one to be banned. He was to be in good company, as Kepler's *Epitome* and Galileo's *Dialogues* were also placed on the odious *Index*. Father Foscarini, a Carmelite, also got in hot holy water with his *Lettera*, a book in which he tried to reconcile certain passages in the Bible with a moving Earth—a very dangerous subject for anyone, let alone a clergyman.

In 1620, the Church printed *Monitum Sacrae Congregationis ad Nicolai Copernici lectorem*. In addition to enumerating what needed to be deleted and changed in Copernicus' book, the document makes the blanket statement that "all other books teaching the same thing" (heliocentrism) were also to be forbidden. As a result, by the middle of the 17th century, the *Index* began to resemble a "Who's Who in Astronomy."

If all this sounds like a foolish practice by a single Pope and a few ignorant Inquisitors, it should be noted that it wasn't until the 1835 edition of the *Index* that the works of Copernicus, Kepler, and Galileo were finally removed from the forbidden list! And the *Index* itself wasn't abolished until 1966!

Fortunately, the Inquisition couldn't reach everyone, nor could it hold the Earth in the center of the universe, no matter how many books it banned or how many people it persecuted. But it certainly wasn't from a lack of trying…

> "The view that the sun stands motionless at the center of the universe is foolish, philosophically false, utterly heretical, because contrary to Holy Scripture. The view that the earth is not the center of the universe and even has a daily rotation is philosophically false, and at least an erroneous belief." Holy Office, Roman Catholic Church, 1616

To See, Or Not to See…

It is difficult to defend a theory when there is no visible proof available. When such proof exists, however, the task of convincing nonbelievers should be childishly simple—unless, of course, those you are trying to convince are childishly obstinate. One of Galileo's first bouts with just such people who refused to take off their blinders was in Pisa in the year 1598.

Aristotle had taught that bodies fall at different rates according to their weight (the heavier the object the faster it falls), although it seems he never bothered to test that theory. Reason was supposed to be superior to experimentation. No one else seems to have tested it either for almost two thousand years, and everyone accepted Aristotelian theories as incontrovertible laws.

Enter Galileo, who did conduct experiments and found that bodies fell at the same rate regardless of their weight. His colleagues at the University of Pisa discounted the idea as being absurd and completely against common sense. According to a popular story, (the authenticity of the story is disputed although Galileo's experiments and the controversies they provoked are certain) Galileo was determined to prove his assertions and utilized the famous Leaning Tower of Pisa to do so.

When students and teachers at the university all gathered at the base of the tower to witness Galileo's experiment, they no doubt thought they were in for a good laugh. However, the joke was to be on them, and Galileo would deliver it with the impact of a lead balloon. Climbing to the top of the tower, Galileo perched two round weights on the edge, one weighing one pound, the other ten pounds (some accounts say the second weight was as much as one hundred pounds).

If Aristotle was correct, the ten pound weight would reach the ground ten times faster. Pushing the two weights over the edge simultaneously, they fell at the same speed and struck the ground at the same instant, their combined sound ringing an end to the Aristotelian world. Or so it should have been.

Rather than thanking Galileo for opening their eyes with this momentous discovery, they refused to believe what they themselves had seen and heard. They became angry with this upstart mathematician, and exerted considerable efforts to try to explain why the experiment was not valid. The persistent ignorance of the faculty and inhabitants of Pisa eventually drove Galileo to resign from the university before his three year post was completed.

By the time Galileo discovered the moons of Jupiter through observations conducted in Padua from January 7 through January 13 of 1610, constant practice had led him to treat the ignorance of his so-called peers with resignation and good humor. Many disputed his claims about the moons, even those who viewed them. Then there were those, including a Professor of Philosophy at the university, who simply *refused to even look* through the telescope! Galileo's reaction to this refusal appears in a letter to Johannes Kepler.

> "Oh, my dear Kepler, how I wish that we could have one hearty laugh together! Here, at Padua, is the principal professor of philosophy whom I have repeatedly and urgently requested to look at the moon and planets through my glass, which he pertinaciously refuses to do. Why are you not here? What shouts of laughter we should have at this glorious folly!"

Those who had argued against the Copernican system had stated that if the Earth moved around the sun, the Moon couldn't possibly stay with the earth. Galileo's discovery of the moons that orbited the moving planet of Jupiter destroyed that argument, and those who obstinately chose to cling to the past had one more reason to hate Galileo. One astronomer in Florence, Francesco Sizzi, refused to believe in the Galilean satellites and stated that raising the number of heavenly bodies from seven to eleven would create nothing less than total havoc.

According to Sizzi, since the seven days of the week were named after the seven heavenly bodies, the addition of four new bodies would ruin the calendar! He also driveled on about the almost magical quality of the number seven, citing examples such as the seven openings in the head and the seven precious metals (of alchemy). Sizzi also stated that, "Moreover, the satellites are invisible to the naked eye, and therefore can have no influence on the earth, and therefore would be useless, and therefore do not exist."

Even one of Kepler's students, Martin Horkey, claimed that Galileo's observations of the moons were nothing more than optical illusions, and that Galileo was interested in money, not truth. Kepler severely chastised Horkey and compelled him to retract those statements. But when even former friends began to criticize his discoveries, Galileo finally lost his good humor and left Padua. Unfortunately, this turned out to be a near-fatal mistake, for when he left the independent Venetian Territory, he moved to Florence—deep in the heart of Inquisition Territory.

> The Bible shows the way to go to heaven, not the way the heavens go.
>
> Galileo

Insult and Injury

Today, there are many 12 Step Programs for treating addictions and numerous personal problems. While such programs boast an encouraging success rate, they pale in comparison with the remarkable success of the old 5 Step Program—sponsored by the Inquisition and guaranteed to get results, or else.

One of the greatest crimes against science, reason, and humanity was perpetrated upon Galileo by the Inquisition, forcing him to recant his heretical heliocentric beliefs put forth in his *Dialogues on the Ptolemaic and Copernican Systems*. The work gingerly suggests that the Copernican system makes for a good hypothesis, but never concludes which system is correct. But even this was enough ammunition for his enemies, who were able to stir up sufficient trouble to have Galileo summoned to Rome.

A portrait of Galileo
by Justus Sustermans, 1636.

(It should be made clear that unlike Giordano Bruno, who was burned for his heliocentric beliefs and criticisms of the Church, Galileo was a devout Catholic his entire life, and despite his advanced age, when summoned, he dutifully obeyed. His downfall, in the eyes of the Inquisitors was not that that he didn't love the Church, but that this love was perverted by his love of science and truth, and in those days there could be only one winner.)

The five steps for extracting a recantation were simple and terrifyingly effective. The first step involved bringing the accused (i.e., the obviously guilty) heretic to the court of the Inquisition and threatening him. If that didn't work, the heretic was brought to the door of the torture chamber and threatened again. After being allowed to think about his possible fate for a while, step three found the heretic being brought inside the torture chamber for a grand tour of all the instruments and devices. Explicit details of how they functioned were no doubt carefully explained.

If the stubborn heretic still failed to see the light of truth, he was then stripped and tied to the rack during step four. The final step involved what the Inquisition termed the "rigorous examination," i.e., inflicting so much pain the poor victim would say or sign anything. (Whatever happened to "Do unto others..."?)

This is what one of the most brilliant men in the history of the world faced for five agonizing months in the year 1633. At the age of seventy, Galileo was imprisoned in Rome and constantly threatened with step five. It's remarkable that his will and health held out that long. When, after these many months, mere threats failed to work, Galileo was brought to the chambers of the Inquisition where no one heard word of him for three days.

Then on June 24, 1633, Galileo emerged with a signed statement which read that he would "altogether abandon the false opinions that the sun is the center" of the universe and that with "a sincere heart and unfeigned faith, I abjure, curse, and detest the said errors and heresies, and generally every other error and sect contrary to the Holy Church."

There is some dispute as to whether Galileo was actually tortured on the rack. He was sworn to secrecy and never spoke of what transpired during those three days. One would like to believe that the Inquisition would not torture an old man for his scientific theories, but let's be realistic—it would have hurt Galileo a lot more than it would have hurt them! Some historians point out that after his "examination," Galileo suffered from a hernia, one of the classic results of being stretched on the rack. But regardless of whether the torture was physical or mental, the effects were nonetheless agonizing.

Despite the horrors he endured and the house arrest under which he would spend the remaining eight years of his life, Galileo's mind could not be imprisoned. He continued to work, not only discovering the libration of the Moon, but writing a work on his laws of motion, *Discourses on Two New Sciences, of Mechanics and of Motions;* arguably his greatest achievement.

After suffering years of illness and blindness, Galileo died at the age of seventy-eight. As a final insult, the Inquisition wanted to deny him burial. Afraid, no doubt, that the truth might rise from the grave and expose their cruel ignorance for all eternity.

They Didn't Call Them Dark For Nothing

After the fall of the Roman Empire, there was an intellectually dim period of history that certainly didn't give astronomers anything to write home about. With the wisdom of the ancient Greeks and Arabs not scheduled to reemerge for centuries, would-be scholars of the Dark Ages turned to the Father of Tabloid Science, Pliny the Elder. Born in the year 23 A.D., Pliny's only surviving work is his *Historia Naturalis*, a lengthy work composed of volumes of virtual nonsense.

He recorded for posterity the "fact" that horses can become pregnant by the west wind, spitting into a snake's mouth will kill it, and, to his credit, that intercourse is a cure for almost every ailment. In addition to reporting about such things as spontaneous sex changes and animals displaying human abilities, Pliny focused his keen eye on the universe. Within the pages of his great monument to science, this tireless seeker of knowledge declares that the universe is essentially a confused mess, obeying no mathematical laws, and completely beyond our ability to ever measure. And what did the mysteries of the Earth, stars, and planets matter anyway; it was obvious that God had no interest in them, so why should we?

Riding the tidal wave of Pliny's intellectual stimulus, generations of Europeans huddled in fear of comets, the aurora borealis, and eclipses. Those few who had some spark of curiosity used it to try to interpret the terrible meanings behind these obvious portents of doom. And let's face it, just about any time in the Dark Ages you predicted disease, famine, or bloodshed, you were probably going to be right.

Six hundred years after Pliny, people like Bede and Gregory of Tours were still mired in the ignorance of their predecessor, maintaining that the heavenly bodies were of no real importance, and still firmly believing in magic and omens. Isidore, Bishop of Seville, in an amazingly feeble and pointless attempt at establishing a "scientific" cosmology, declared that the entire universe was a

symbolic representation the church. How many nights did he burn the midnight oil calculating that one?

Unfortunately, such ideas did not begin to face scientific or mathematical scrutiny until the twelfth century, when ancient texts began to be translated into Latin. It is truly an astonishing feat that for over a thousand years, no serious observational data was compiled, no useful instrumentation was developed, and the collective Western intellect was content to sit in the murky backwaters of superstition.

However, it was the fresh ocean waters that would ultimately snap Europe out of the doldrums. When the importance of the celestial realm for accurate navigation was recognized, the scientific ranks sprang to attention—lured not by the invigorating scent of salt air, but by the seductive smell of commerce.

Can't Make Them Think

By the year 1800, the persecution of astronomers and much of the fear and ignorance surrounding the science was finally a thing of the past. While great strides were being made in all areas, however, some people chose to stick with the past, and there are those who still do to this day.

Since the time of the ancient Greeks, there have always been those who believed our world was round, not flat. Many times throughout history such people were in the minority, but one would think that by the beginning of the 19th century, the issue would have been settled once and for all. Not so. Certainly in our present age, when astronauts have taken pictures of our plump, spinning Earth, there can't be anyone who still clings to the concept of a flat world. Not so, again.

In the early 1800s, a society was formed in the United States and England that not only maintained that the Earth was flat, but that the sun revolved around us. The North Pole is at the center of this flat Earth and the continents and oceans stretch out around it. Framing the world is Antarctica; not a separate continent surrounded by water as was generally believed, but running along the outermost edges of the Earth in a long, frozen strip.

Early in the 20th century, when explorers like Amundsen, Scott, and Byrd claimed to have reached the South Pole or crossed the continent, members of the Flat Earth Society (then known by its original name, the Zetetic Society) declared that the explorers had been confused and had merely traveled along the perimeter of the Earth.

What of the Apollo Moon landings? A hoax, of course, Flat Earthers contend. How could anything NASA says be true when it is clear that the Moon is only 32 miles wide and no more than 2,500 miles away? As for the sun, it is a little larger than the Moon and orbits us only 700 miles beyond the orbit of the Moon. So despite an overwhelming body of scientific evidence, including countless photographs, there are still those today, who only see exactly what they want to see.

What can be concluded from the Flat Earthers and their adamant beliefs? You can lead a person to knowledge, but you can't make him think.

"I think there is a world market for maybe five computers."

Thomas Watson, chairman of IBM, 1943

"Computers in the future may weigh no more than 1.5 tons."

Popular Mechanics, 1949

Six and Six Equal Nothing

Christiaan Huygens' name would have been remembered for any one of his many accomplishments—the wave theory of light, the pendulum clock, and improvements in telescopes and eyepieces. He was also the first to explain the true nature of Saturn's puzzling structure, the magnificent rings (1656), and discovered the first satellite of Saturn, Titan (1655). Huygens no doubt would have made even more discoveries, had it not been for an odd misconception.

Upon his discovery of Titan, he noted that, together with our Moon and the four moons of Jupiter, it brought the total number of

satellites in the solar system to six. Six also happened to be the number of known planets.

Coincidence? Huygens didn't think so. In fact, he was convinced that there was some great significance in this apparent equal number of planets and moons. Firmly believing that there was no possibility of discovering any more moons (because their number *obviously* couldn't exceed the number of planets), he essentially stopped looking for them.

Fortunately, Huygens' contemporary, J.D. Cassini, was not harboring such unfathomable, preconceived notions, and went on to discover four more satellites of Saturn—Iapetus 1671, Rhea 1672, and Tethys and Dione 1684. Huygens, the great mathematician, must have been stunned to discover that six and six actually equal nothing.

Rising Above The Occasion

The German military machine that plunged the planet into two world wars got itself rolling during its victories in the Franco-Prussian war of 1870-71. War machines tend to indiscriminately roll over everything in their path, but occasionally a clever individual can rise above the turbulence.

One such individual was Jules Janssen, founder of the Meudon Observatory in Paris and a leader in the field of solar research. Prior to 1868, astronomers were limited to viewing prominences and the chromosphere only during a solar eclipse, but while on a trip to India (to observe a total eclipse), Janssen discovered a method for observing the outer regions of the sun without the aid of the Moon (Sir Joseph Lockyer made a simultaneous discovery in England).

Janssen was obviously not one to let distance get in the way of his research, and old age and weather proved to be no obstacles, either. During the winter of 1894, the seventy-year-old Janssen was carried up Mont Blanc in the Alps, hoping that the thin atmosphere would enable him to determine if there was any oxygen in the sun. Even war could not daunt the tenacious astronomer.

During the Franco-Prussian War, the enemies of France were rude enough to put Paris under siege, not only disregarding the inhabitants' life and liberty, but their pursuit of scientific knowledge, as well. The problem for Janssen was that he was stuck inside the city and the path of totality for an upcoming solar eclipse fell outside the line of the siege. Apparently, the enemy ranks were not filled with astronomy enthusiasts and the determined Janssen decided that if he couldn't go through them, he would go over them.

Circa 1920 bas relief at Meudon of Janssen's ascent.

In a brilliant and courageous triumph of Good Astronomer over Bad Astronomy Besiegers, Janssen made good his escape from Paris in 1870 in a hot air balloon. Unfortunately, it is mainly for this exploit and not his discoveries that he is best known, but who could object to being remembered for jeopardizing one's personal safety in the name of science? We should all be so lucky.

Send in the Clowns

Our "knowledge of the construction of the Moon leads us insensibly to several consequences...such as the great possibility, not to say the almost absolute certainty, of her being inhabited." (The key word here, being "insensibly.")

These were not the ravings of a crackpot, they were the words of astronomer Sir William Herschel, and were spoken to his colleagues in 1780. What was not said out loud were some of the reasons why Herschel so firmly believed in life on the Moon.

In his private journals, Herschel recorded seeing many strange phenomena, much like the TLP (transient lunar phenomenon) still observed today. It is understandable that such sudden appearances of

glowing lights could be interpreted as the result of some action by intelligent life forms. What is incomprehensible, however, is what else Herschel reported seeing.

Herschel, one of the greatest observers in the history of astronomy, thought he saw lunar towns, forests, and highways. He also recorded in his journal the word "Circus." However, he was referring to the Roman term for something circular, not the clown variety—although given everything else he believed he might as well have meant a lunar circus!

For example, Herschel wrote:

"As upon the Earth several Alterations have been, and are daily, made of a size sufficient to be seen by the inhabitants of the Moon, such as building Towns, cutting canals for Navigation, making turnpike roads &c: may we not expect something of a similar Nature on the Moon? – There is a reason to be assigned for circular-Buildings on the Moon, which is that, as the Atmosphere there is much rarer than ours and of consequence not so capable of refracting and (by means of clouds shining therein) reflecting the light of the sun, it is natural enough to suppose that a Circus will remedy this deficiency, For in that shape of Building one half will have the directed light and the other half the reflected light of the Sun. Perhaps, then on the Moon every town is one very large Circus?...Should this be true ought we not to watch the erection of any new small Circus as the Lunarians may the Building of a new Town on the Earth....By reflecting a little on the subject I am almost convinced that those numberless small Circuses we see on the Moon are the works of the Lunarians and may be called their Towns....Now if we could discover any new erection it is evident an exact list of those Towns that are already built will be necessary. But this is no easy undertaking to make out, and will require the observation of many a careful Astronomer and the most capital Instruments that can be had. However this is what I will begin."

Fortunately for Herschel and his family, these private records remained private until relatively recently.

Ironically, it was Sir William's son, Sir John, whose name would be used for the great Moonman hoax of 1835 (see below). Right story, wrong Herschel.

Locke, Stock, and Barrel

In 1938, thousands of sane and rational individuals huddled in fear around their radios as Orson Welles told of Martians invading New Jersey. While the presentation of "War of the Worlds" plunged the East Coast into a panic, the fame and notoriety from it propelled Orson Welles to the West Coast—Hollywood to be exact, where the rest is cinematic history. Although he was obviously pleased with the results of his little radio drama, he was quite surprised by the uproar it provoked. Authorities were flooded with calls, men grabbed their guns, and women and children anxiously searched the skies for some sign of the terrible invaders.

There is probably some kind of psychological theory which suggests that people will only believe something so outrageous if they have some deep-seated need to believe it. In the case of extraterrestrials, mankind has longed for proof of their existence for centuries, and therefore, many people are ready and willing to suspend their disbelief in the brief instant it takes a strange light to flash across the night sky.

There's absolutely nothing wrong with the desire to find some neighbors in this vast universe. However, before the public swallows another story like the Martians in New Jersey, it should make sure the bait isn't rotten.

One of the biggest alien fish stories ever swallowed actually occurred over one hundred years before Orson Welles' ominous voice terrified his listeners. In 1835, a failing newspaper, the *New York Sun*, had an explosive boost in circulation thanks to some creative journalism—i.e., a hoax. An enterprising reporter for the *Sun*, Robert Locke, apparently took it upon himself to single-handedly save the paper. Taking the liberty of using astronomer Sir John Herschel's good name and family reputation, Locke fabricated a story too unbelievable not to be believed.

Claiming to have exclusive access to a paper Herschel had supposedly published in a Scottish journal (which had actually been defunct for two years), Locke reported that Herschel had discovered an incredible civilization of yellowish, bat-like humanoids on the Moon. In addition to this intelligent race of winged beings, the story claimed that 130 species of lunar plants and animals had meticulously been catalogued. This was all allegedly made possible by the incredible telescope Herschel was using, whose 42,000-power optics could capture such amazing detail.

The winged Lunarians, from the *New York Sun*, 1835.

Impossible? Ridiculous? Of course, but the public ate it up like candy. Tens of thousands of newspapers and pamphlets sold in a matter of days and practically everyone was caught up in Lunarian Fever. Some cooler heads did prevail, however, as a few scientists from Yale tracked down Locke and pressured him into admitting to the hoax, but not before many respected people made some wild plans for the flying Moonmen. One of the most bizarre plans was to Christianize our happy, but heathen, neighbors, although the little matter of how these preachers and their Bibles would get to the Moon wasn't completely explained.

While such hoaxes in this age of rapid communication are harder to perpetrate (Herschel was not informed of the hoax for months), our increasing angst about finding other life in the universe may make us vulnerable to the schemes of some knowledgeable and clever huckster. One need only look at the tabloid newspapers at every grocery checkout to see the high degree of nonsense some people are willing to swallow—Locke, stock and barrel.

Castles in the Sky

One of the first astronomers to correctly suggest that the craters on the Moon were formed by impacts was Franz von Paula Gruithuisen in 1824. He had arrived at this theory after extensive observations—observations that occasionally took on a rather creative dimension.

In 1822, Gruithuisen reported on another alleged aspect of Earth's satellite—nothing less than a city on the Moon. He believed he had observed "great artificial works on the moon erected by the lunarians." This city was supposedly protected by a series of ramparts and fortifications constructed by "selenic engineers."

However, the "dark gigantic" structures Gruithuisen claimed were the design of lunar architects were, in fact, nothing more than a series of small, irregular natural ridges that had been reconstructed by his imagination. Despite the reports of these castles in the sky, Gruithuisen was made a Professor of Astronomy in Munich in 1826 and spent many years making serious contributions to science.

(Perhaps the least known of his contributions is the "Gruithuisen Effect," which is still observed quite often. This phenomenon involves making outrageous mistakes and then getting promoted—an indispensable process in both the corporate and academic worlds today.)

As The Planets Turn

The history of an astronomical discovery often reads like a great mystery novel. Occasionally, it more closely resembles a soap opera. The search for Neptune had mystery, soap, and more:

1. Where's Uranus?— Sir William Herschel discovers Uranus in March of 1781. Other astronomers had observed the planet before him, but failed to notice that this faint "star" moved. Herschel's

career skyrockets; the other astronomers enter the dark realm of footnotes.

2. Now Where's Uranus?— For several decades no one is able to calculate Uranus' orbit. Because the planet is never where it's supposed to be, some theorize that an eighth planet is perturbing Uranus' orbit.

3. On With The Hunt— The young, brilliant Englishman, John Couch Adams, begins his calculations in the search for the new planet in 1843. He succeeds, on paper, in September of 1845.

4. What's The Rush?— Adams delivers his calculations to George Airy, the Astronomer Royal, and James Challis, the Professor of Astronomy at Cambridge. The two astronomers and their great observatories (Greenwich and Cambridge) apparently have no interest in discovering a new planet and do nothing.

5. Enter The Frenchman—While the English twiddle their thumbs, Urbain Leverrier (unaware of Adam's work) publishes his own calculations on June 1, 1846. However, the French telescopes don't start searching either, most likely because over the years Leverrier has irritated just about everyone in France.

6. Duh?— Airy reads Leverrier's work and finds that the results of the Frenchman's calculations match Adams'. Maybe someone should actually start looking?

7. I'll Get Around To It—Airy asks the sluggish Challis to begin the search. After devising the slowest search strategy possible, Challis begins. He actually records the unknown planet on August 4 and then again on August 12, 1846, but doesn't bother to check if the position of this particular "star" had actually changed. Challis also doesn't bother to increase magnification, a simple task which would have revealed that this "star" was indeed a planet.

On September 29, Challis spots something that appears to be a planet, but again doesn't bother to increase the magnification. It can wait.

The next night, Challis spends so much time at dinner that it's cloudy by the time he finally gets his butt moving.

8. German Efficiency—Leverrier gives up on the French and writes to the observatory in Berlin. (Apparently, he hasn't had the opportunity of aggravating them, yet.) They receive his calculations on September 23. Johann Galle and Heinrich d'Arrest begin looking that same night and find the mystery planet. They confirm it by its movement the following evening. Game, set, match to the Germans.

John Couch Adams, and Leverrier's grave at the Cimetière du Montparnasse in Paris.

9. Kick Him Again—Everyone congratulates Leverrier for the discovery, including Airy, who knew damn well that Adams had predicted the new planet's position a year earlier.

10. Son of Herschel To The Rescue—Realizing the ill treatment of Adams, Sir John Herschel flexes some academic muscle and Adams' calculations become known.

11. Now We Like Him—Finally realizing this is their chance to trump both England and Germany, the very same French astronomers who wouldn't search for the planet suddenly decide it should be named "Leverrier."

Leverrier thinks it's a good idea, too.

12. Classicism Wins Out—After much *Sturm und Drang*, the Germans' desire for the planet to be named Neptune is fulfilled.

Epilogue—John Couch Adams went on to become the director of the Cambridge Observatory. The humble and dedicated astronomer actually turned down an offer of knighthood out of concern that the attention would detract from his work.

Leverrier went on to annoy more people in his search for Vulcan, the planet he claimed was between the sun and Mercury. He based his claims on the observations of an inept amateur, Edmond Lescarbault, who took notes on pieces of wood and once mistook Saturn for a new star.

Can't Argue With Success

In most instances, individuals or small groups are responsible for acts of Bad Astronomy. In a few remarkable cases, however, the error is so widespread that the sheer volume of people involved almost overshadows the ridiculous practices or beliefs.

Almost.

One such case involves eclipses—events met with fear around the world for thousands of years. From the dimmest recesses of our past to the beginning of the 20th century, cultures in Asia, Africa, Europe, and the Americas viewed eclipses as the work of some terrible monster attempting to devour the Moon. Regardless of whether the culprit was believed to be a dragon, serpent, demon, dog, or giant, the general consuming scenario remained the same.

Absurdities aside, it is remarkable that so many cultures isolated from one another arrived at such similar beliefs. What is even more startling is how everyone from royalty to peasants, and from the tiny islands in the Pacific to bustling Western cities, arrived at the same remedy for getting the monsters to cough up their celestial meal—noise.

Noise on a deafening scale.

Entire communities would turn out at the first sign of an eclipse and join together to shout, beat drums, and, the popular favorite among old and young alike, banging pots and pans. It was assumed that if the people made enough of a clamor, it would frighten the ravenous beast and cause it to spit out its food and run away. Why a creature large and powerful enough to swallow the Moon would be scared off by a bunch of rowdies armed with skillets remains a mystery.

Joseph Gyscek

This bizarre practice stretched across millennia and the length of the globe for a simple reason—it always seemed to work! Every time there was an eclipse, the people raised a ruckus and the Moon would miraculously reappear unharmed. A one hundred percent success rate is a tough thing to argue against, and it is probable that many a rational objection to this practice was literally silenced beneath the din of popular opinion.

These countless generations of pot-bangers may also be forgiven some of their folly by considering what a tremendous boost to the personal ego and community spirit the defeat of the monster must have brought. In a human history plagued with myriad disasters beyond anyone's control, here was one brief and shining moment where friend and foe alike would band together to drive off a common enemy.

In retrospect, perhaps this is a practice that should be revived.

Blood, Sweat, and Fears

Making loud noises was standard practice in many cultures around the world to ward off the demons or monsters that were believed to cause an eclipse. The Aztecs did their share of screaming and crying, but they also had a few unique reactions to their eclipse fears.

Pregnant women, afraid that evil effects of the eclipse could cause their unborn children to become deformed, would put pieces of obsidian between their breasts to protect them (the unborn children that is, not their breasts). Another concern was that the celestial event could cause women to give birth to rodents, but apparently a well-placed rock was also sufficient to avert that fate.

The Aztec people, never known for being squeamish, also decided that making everyone bleed was a good way to protect the general populace. However, even this practice is tame in comparison to their ultimate response, human sacrifice. If there was ever any bad luck occasioned by an eclipse, it fell upon the hapless prisoners who were ritually murdered by the superstitious Aztecs—not only a very rude practice, but one of the worst cases of Bad Astronomy ever recorded.

You Think Your Job Is Tough?

(The following is most likely only a legend, but most legends have a kernel of truth and this one was too tasty not to be popped.)

The ancient Chinese were known for their meticulous astronomical observations, yet were by no means free from the superstitions that permeated less educated and disciplined cultures. Eclipses were events to be feared and accurately predicting their occurrence was imperative so that the learned men could immediately begin conducting the ceremonies designed to stave off any evil effects. (It was common knowledge that high officials burning incense and candles, while striking gongs and praying, were always able to bring back the sun or Moon. Talk about job security!)

Such predictions were part of the job description of Hi and Ho, two court astronomers who imprudently spent too much time with the wrong kind of moonshine. Failing to predict an eclipse because of their drunkenness, the Emperor was forced to mete out stiff disciplinary action, as any conscientious employer would have done. Indeed, the punishment was most effective, for after beheading both negligent astronomers, they never made another mistake.

Eclipsing Reason

The Peloponnesian War was fought between Athens and Sparta during the years 431-404 BC. The story of the war contains rousing accounts of bravery and treachery, tragic tales of plague and starvation, and a single episode of Bad Astronomy in which the Moon inadvertently turned the tide of war—and possibly the course of all of Greek history.

A peace agreement of 421 BC was broken when the clever, but overly ambitious, Alcibiades persuaded his fellow Athenians to attack Syracuse. After the fleet was launched, Alcibiades' enemies made some false accusations that were disturbing enough to have him removed from command—a command which fatefully fell into the superstitious and incapable hands of Nicias.

While Alcibiades quietly slipped away and joined forces with the Spartans, Nicias was getting the Athenian fleet into deep trouble. It seems that in August of 413 BC, the inept Athenian general was about to become trapped by the Spartan fleet in the harbor of Syracuse. There appeared to be a good opportunity for Nicias to escape the trap and save his fleet, were it not for a dreaded lunar eclipse. Terrified by the omen, Nicias consulted his astrologers who advised him to stay where he was—for twenty-seven more days! Despite a lack of food and a strategically suicidal position, Nicias abandoned reason, took their advice, and refused to set sail.

Apparently, the Spartans weren't the least bit intimidated by the celestial event, and promptly on the following day, they surrounded the Athenian forces and destroyed them. The battle proved disastrous to Athens (half their fighting force was effectively lost), who after gallantly struggling for several more years, ultimately surrendered to Sparta.

As for Nicias, the eclipse became a self-fulfilling prophesy, as he was captured in the battle and executed. The Athenian captives perhaps fared even worse. They were sent to work in the mines—a virtual sentence of death. On the bright side, the Athenian defeat encouraged twenty thousand slaves working in Athens' mines to revolt and gain their freedom. Not surprisingly, the irrepressible Alcibiades continued switching sides several more times, and was finally assassinated in 404 BC.

It is impossible to say that the outcome of the war would have been different had Nicias not been immobilized by the shadow of the earth—chances are he would have eventually done something else stupid which would have led to defeat. However, this episode does prove that while a little knowledge may be a dangerous thing, a complete lack of knowledge can be fatal.

Now You See It, Now You Don't

(In all fairness, the following is not technically a case of Bad Astronomy. However, it is so unethical that it has earned the right to be dishonorably mentioned. Let your conscience be your guide.)

During Columbus' four voyages to the New World, the natives generally treated him and his men with kindness and generosity. In return, when the Spaniards began grumbling about living conditions and the lack of gold on the Caribbean islands, Columbus pacified them with a few perks—namely giving them the natives' land and throwing in the natives themselves to work as slaves. Occasionally, some of the local inhabitants failed to appreciate the honor of serving their new masters, but Spanish steel and a volley of hot lead often resolved the misunderstanding.

Columbus, dazzling the natives with a bell.

On May 9, 1502, Columbus embarked on his final voyage, hoping to recoup the prestige that had been so tarnished during the decade since his first voyage. Convinced that South America was only a stone's throw away from China, he sailed his men up and down the coast of Central America looking for a passage to the Far East.

Conditions were horrific. Not only were worms eating his ships, but as his son, Ferdinand, related, "What with the heat and wet, our hardtack became so wormy that, God help me, I saw many sailors who waited till darkness to eat it so they would not see the maggots."

The ships became so badly damaged and waterlogged they could go no further than Jamaica. While waiting almost a year for new ships, Columbus and his men once again treated the natives in their time-honored, high-handed manner. The ungrateful natives, failing to recognize the innate nobility and superiority of the Europeans, became quite peeved and refused to supply any more food. Faced with a hundred starving, mutinous men, Columbus resorted to the oldest trick in the astronomer-priest book.

Picture a beach in Jamaica in the early evening of February 29, 1504. Donning his finest suit of clothes (probably the one with the least amount of worm holes), Columbus proclaimed to the natives that God was not happy with their lack of hospitality, and was going to take the Moon away to punish them. At first skeptical, few natives gathered to watch the rising Moon. However, it quickly turned an ominous, blood red, and then began to slowly disappear. Suitably intimidated by the awesome power of the European Admiral, the locals rushed baskets of food to the Spaniards, begging for forgiveness and asking them to intercede with God and return the Moon.

Columbus coolly replied he would think about it, and let the terrified natives sweat for a while. Finally, to their great relief, he agreed to ask God to bring the Moon back, and miraculously it was restored, right on schedule.

Everyone was happy—the Spaniards got their food, Columbus got back home, and the natives of the New World eventually met virtual extinction through foreign diseases such as small pox, but were better people for having known the Europeans.

Some may argue that Columbus was clever and had every right to use the knowledge of the impending eclipse to his advantage. After all, why be nice to people when you can use information and technology to scare the hell out them instead?

While Columbus cannot really be convicted of Bad Astronomy for this Jamaican shell game, it is doubtful that he will ever be posthumously granted any Advancement of Science or humanitarian awards for the event.

(Note: The book from which Columbus obtained his eclipse information was *Ephemerides* written by Regiomontanus in 1474. The book contained many errors, but fortunately for Columbus, the date and time of this eclipse was accurate.)

Aiming To Please

William Herschel began his career as a musician and therefore knew how to please an audience. This training was to come in handy when he was summoned to the court of King George III of England.

During one of Herschel's four "sweeps" of the sky (each of which took years as he laboriously catalogued every object), he came across something new near Eta Geminorum on March 13, 1781. He reported it to be a comet, but after other astronomers studied the object, it was identified as a new planet. Herschel's discovery of Uranus was monumental—since antiquity, only five planets had been known and Europe's waning interest in astronomy was instantly reignited.

Herschel was catapulted out of the ranks of amateurs, into the Royal Society, and the rarefied atmosphere of the court. The king was most anxious to meet the celebrated astronomer and look through his telescope, so Herschel packed up his best one and went off to Windsor Castle.

Lords and Ladies flocked to see the wonders of the heavens and Herschel planned a gala observing night in hopes of satisfying them all at once. Of course, on the planned night the unwritten Law of Observation was in effect—the one stating that the quality of weather is inversely proportional to the importance of the viewing event. Mere clouds and the threat of rain wouldn't stop Herschel from pleasing the royals, however, and he arrived upon an ingenious solution.

No doubt after sizing up the intellectual capacity of his audience, Herschel decided that if they would be unable to see the real planet Saturn, he would create his own. Making a mock-up out of paper, he hung it at some distance in the garden and illuminated it from behind with a lamp. It was upon this pseudo-Saturn that Herschel aimed his telescope, much to the delight of his dim, yet regal, audience. Apparently, you actually *can* please all of the people some of the time.

Joseph Gyscek

Et Tu, Augustus?

Julius Caesar was stabbed to death in the senate by a group of his friends and allies on March 15, 44 BC. In September of the following year, a comet was spotted over Rome and immediately interpreted to be the soul of the murdered hero ascending to heaven. Unfortunately, records give no explanation as to why Caesar's soul put off the spectacular journey for 18 months, but who are we mere mortals to question such things?

There is another slight glitch in the story. The meticulous Chinese astronomers did note a comet in May-June of 43 BC, but none in September. The Romans claimed that this auspicious "star of the mightiest Julius" would make its appearance around 4pm and remained visible for five to seven days. It is unlikely that such conscientious observers as the Chinese could have missed a comet brilliant enough to be seen in daylight for a week, but they may have just been jealous since none of their emperors had thought about making such a grand entrance into heaven.

Twenty-five years later, Caesar's comet/soul was still fresh in the memory of his grandnephew, Emperor Augustus (who had been Caesar's sole heir, thereby ensuring a healthy memory). Augustus was holding games called the Saeculares, when lo and behold, another comet appeared. Anybody with half a Roman brain obviously realized that this was Caesar's soul returning to proclaim what a great emperor Augustus was.

Associating yourself with the comet/soul of a murdered legend makes for good press in any age, and Augustus promptly ordered silver coins to be struck commemorating the divine event. The coins, bearing the portrait of Augustus and a representation of Caesar's comet/soul, remain as a witness to the fact that while some things are clearly Bad Astronomy, they nonetheless are Good Politics.

The coin bearing Caesar's comet/soul.
American Numismatic Society

Good Comet, Bad Comet

When it comes to war, one man's evil omen is another's divine blessing. Such was the case with the appearance of Comet Halley in the year 1066. For Harold, King of England, the comet signaled a threat to his country, his reign, and his life. To William, Duke of Normandy, it appeared as a messenger from the heavens affirming that his cause (the invasion of England) was just. Obviously, there was a lot more going on there than just a comet sighting.

To briefly summarize one of the most pivotal events in English history:

1. Harold's predecessor, Edward the Confessor, spends thirty years in Normandy before becoming King of England. In gratitude for the protection afforded Edward by the Norman nobles during his lengthy stay, he promises to make William heir to the throne of England.

2. In his pre-king days, Harold is knighted by William, and in the process swears his allegiance and promises to support William's claim to the throne once Edward passes away.

3. Edward dies on January 5, 1066. Harold grabs the throne and becomes king. William becomes royally miffed.

4. On April 24, Comet Halley appears and is visible for a week. Harold is frightened by the "long haired star." The Normans think it's a sign that the lying Harold is going to get what's coming to him.

5. As the Normans prepare to invade England, Harold's exiled brother Tosti jumps on the bandwagon and sides against his regal sibling. While Harold's army defeats Tosti's army at Stamford Bridge on September 25, the victory is costly.

6. William lands 1400 vessels on the English coast. The Norman army meets Harold's weakened forces near Hastings. On October 14, during a battle that lasts all day, Harold gets what's coming to him. After being blinded by an arrow in the eye, he is promptly hacked to pieces. The English flee and William earns the title "The Conqueror." He is crowned king on Christmas Day.

To commemorate the conquest, the famous Bayeux Tapestry was created. The impressive tapestry (which is an astounding 230 feet long) contains among its sixty scenes a depiction of the fateful comet. In this scene, Harold looks so shaken that it appears he's about to fall off his throne. Members of his court point up at the celestial spectacle and look suitably terrified, humbled, and ready to be conquered.

While it is possible this depiction is accurate, it must be remembered that the chronicling of history is often more of a political art than a science. Had William hidden under his bed at the

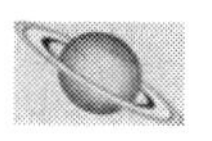

sight of the comet, it is guaranteed that not a single thread of a tapestry would have shown it. That's one of the perks of being a conqueror.

The famous scene in the Bayeux Tapestry where the comet appears as a portent of doom to Harold.

In The Eyes of the Beholder

In terms of celestial events evoking sheer terror, it's probably a dead heat between eclipses and comets. However, while eclipses are more frequent, they are relatively brief. Comets on the other hand, appear infrequently, but can last for days or even weeks. Throughout history, the prolonged agony of a comet apparition has actually proven to be too much for some people to bear.

The epitome of comet-fear is wonderfully expressed in the following eyewitness account of the comet of 1528:

It was "so horrible and produced such great terror in the common people that some died of fear and others fell sick. It appeared to be of excessive length and was the color of blood. At its summit rose the figure of a bent arm, holding in its hand a great scimitar as if about to strike. Three stars quivered near the tip of the blade. On both sides of the rays of this comet there appeared a great number of axes, knives and blood-drenched swords, among which were many hideous faces with beards and bristling hair."

Damn, they just don't make comets like they used to.

Note: The eyewitness of this comet was Ambrose Pare, a noted surgeon. His specialty was treating a relatively new kind of injury—the gunshot wound. One of his innovations was to discard the Arab treatment—namely cauterizing the wound with boiling oil or hot irons—and replaced it with ligature. As his native France seemed perpetually embroiled in war, Pare unfortunately had plenty of opportunities to test new treatments. After the unceasing horrors he witnessed, it is no wonder he would view a comet with dread!

It's the "Thought Balls" that Count

For thousands of years, comets have been accused of bringing death and destruction to Earth, but never before had they been accused of bringing aliens. Well actually, it wasn't that the aliens were supposedly riding on the comet; they were just tagging along in their enormous planet/spacecraft to bring enlightenment to the people of Earth. But first things first.

Comet Hale-Bopp was discovered on the night of July 23, 1995. Early predictions called for it to be one of the brightest comets of the century, but the public had heard that one before. (Astronomers' claims such as, "This is going to be the mother of all bright comets," were in peril of joining the ranks of such statements as, "The check is in the mail," and "Of course I'll respect you in the morning.") Skeptics worried that Hale-Bopp would only last as long as a politician's campaign promises, but fortunately for astronomers, it was indeed a brilliant and spectacular comet.

For over a year, professionals and amateurs observed the comet as it slowly moved through the solar system, growing ever-brighter as it approached the sun. Magnificent photographs and CCD images were being taken by the thousands from every corner of the globe,

but none caused as much of a stir as those taken by an amateur in Houston, Texas. Known primarily for its oil, college football, and cow-slaughtering, Texas quickly became the focal point for an alien invasion on a scale surpassing even that of the Mexican border.

The spectacular twin-tailed Comet Hale-Bopp.
E. Kolmhofer, H. Raab; Johannes-Kepler-Observatory, Linz, Austria

From his Houston home on the evening of November 14, 1996, amateur astronomer Chuck Shramek took a series of CCD images of Hale-Bopp through his 10-inch telescope and noticed an "odd thing" near the comet. It appeared to be star-like with small spikes of light, causing Mr. Shramek to refer to it as a "Saturn-like object," or SLO, as it quickly came to be known. And it came to be known so quickly, because rather than diligently investigating and confirming that the object was something new in the heavens, Shramek went on the radio that same night to announce his discovery. Had he properly used his computer star catalogue, he would have immediately identified the alleged SLO as the ordinary star SAO141894. But then again, had he done that, we all would have missed the fun which was about to begin.

The radio program on which Shramek made his announcement was the *Art Bell Show*, a syndicated program devoted to the paranormal, and what started as a mistake of astronomy, frantically spiraled into an astronomical mistake. Personages of the likes of author Whitley "Please abduct me again so I can have another bestseller" Strieber, and Professor Courtney Brown of the dubious Farsight Institute jumped on the bandwagon and proclaimed that the SLO was actually a massive alien spacecraft.

They reached this conclusion because Professor Brown had assigned a team of the Institute's psychic "remote viewers" to scan the object with their telepathic abilities. (In the interest of accuracy, it should be noted that these psychics refer to themselves as "Scientific Remote Viewers," obviously to distinguish themselves from the unscientific ones.)

These remote viewers described a half-planet, half-spaceship object filled with an alien race just itching to bestow its wisdom on the less fortunate races, e.g., humans. And even more remarkable, was the fact that these aliens did not think in a primitive linear manner as we do. They think in "thought balls." And what red-blooded human has not fantasized about having a good alien thought ball?

Voices of reason did cry out in this wilderness of Bad Science. Battlegrounds sprang up, primarily on the Internet, as astronomers armed with the facts attempted to shoot down the wild theories which snowballed around the world. But what mere scientific proof has a chance against those who claimed that even the Vatican was involved in a conspiracy to hide the truth of the approaching aliens?

During the spring of 1997, as Comet Hale-Bopp completed the inner-system tour of its orbit, millions of people around the world were treated to sensational views of a rare and beautiful celestial event. Were the pro-alienists disappointed that a huge spacecraft was not part of the spectacle? No doubt. But if we are lucky, by the time Hale-Bopp returns in another 4,000 years, mankind will have evolved to the point where we can build ships to follow comets and carry our own thought balls throughout the universe.

> "A rocket will never be able to leave the Earth's atmosphere."
>
> New York Times, 1936

All Dressed Up and Look Out Below!

The ancient Romans recognized that nothing enhances the prestige of a cult or religion like a solid piece of divinity to display before the public, and nothing could have been more solid or more divine than a chunk of rock hurled down from heaven by the gods themselves.

A meteor streaking across the sky and slamming into the ground was interpreted as an actual image of a god descending to Earth. Therefore, the Romans considered meteorites to be sacred. Temples were erected across the ancient world to house these sacred stones, and they were worshipped as a manifestation of a god. One of the most famous was the image of Artemis at Ephesus, which is mentioned in the Bible.

"Men of Ephesus, what man is there that does not know that the city of the Ephesians is temple keeper to the great Artemis and of the sacred stone that fell from the sky?" (Acts, XIX, 35) There is an earlier biblical reference possibly referring to meteorites in the book of Joshua, when, "the Lord threw down great stones from heaven," (X, ll) to slay an enemy, but no one seemed interested in preserving or building temples for these holy projectiles. These lucky children of the Almighty simply contented themselves with having their enemies crushed to bloody pulps.

In Greek and Roman texts, there are references to four separate meteorites, although there seem to have been numerous temples claiming to house sacred stones. This leads one to believe that, like medieval relics of saints that often turned out to be nothing more than animal bones, not all sacred stones were of divine origin.

Another meteorite was the sacred black stone at the temple of the Sun God in Emesa (Syria) and it had a rather colorful journey through the ancient world, one which actually rivaled its trip through the heavens to Earth. A young priest at the temple was so enamored of the meteorite that when he left Syria he took it with him. The boy, born Varius Avitus, was commonly known as Elagabalus, after the name of the god he worshipped. Suddenly finding himself Emperor, Elagabalus took the stone (also known as the monolith of Baal) to Rome, and rather than discreetly introduce his native sun worship to his conservative subjects, proceeded to build a huge temple for the meteorite on the Palatine hill.

As a further display of his reverence, Elagabalus dressed up his meteorite, put it in a chariot, and personally led an odd religious procession through the streets every year. The unusual event is described in the fifth book of the histories of Herodian:

"He placed the Sun-god in a chariot adorned with gold and jewels and brought him out from the city to the suburbs. A six-horse chariot bore the Sun-god, the horses huge and flawlessly white, with expensive gold fittings and rich ornaments. No one held the reins, and no one rode in the chariot; the vehicle was escorted as if the Sun-god himself was the charioteer. Elagabalus ran backwards in front of the chariot, facing the god and holding the horses' reins. He made the entire journey in reverse fashion, looking up into the face of his god. Since he was unable to see where he was going, his route was paved with gold dust to keep him from stumbling and falling, and bodyguards supported him on each side to protect him from injury."

Elagabalus and his meteorite in a chariot are depicted on this ancient Roman coin. American Numismatic Society

While this strange religious procession irritated his Praetorian Guard, it was tolerated. However, many other activities of the insatiable Elagabalus were not, and he was assassinated in a latrine when he was eighteen. Unlike its unfortunate admirer, the holy meteorite continued to be revered and probably made its way back to its original temple in Emesa to be worshipped by succeeding generations.

 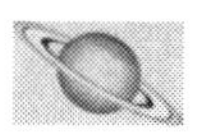

Today, meteorites continue to be coveted, with the only difference now being that science has taken a back seat to the Money-god.

"I would sooner believe that two Yankee professors lied, than that stones fell from the sky."

Thomas Jefferson, regarding reports of meteorites in the 1790s

Divine Protection?

Since the Turks had overthrown the Byzantine Empire and conquered Constantinople in 1453, Europe was a bit on edge as the upstart Ottoman Empire looked to acquire even more real estate.

On November 16, 1492, a meteor slammed into a wheat field in the Alsatian town of Ensisheim. The Holy Roman Emperor, Maximilian I (best known for his effective use of the interchangeable and equally hazardous tactics of marriage and war) declared that the meteorite was a sign from God of divine protection from the advancing Turks.

A document from 1492 depicting the fall of the meteorite from heaven.

The holy meteorite was then brought into the church. The clever leaders of the day even thought to chain the meteorite to the church floor—not to foil thieves, but to keep the symbol of holy protection from flying back into the sky! (Don't you hate when that happens?)

The meteorite apparently did its job. Not only did the Turks fail to conquer Europe, but much of the continent became the inheritance of Maximilian's heirs.

So, if you are ever in need of divine meteoric protection, this historic stone is still on display at the Ensisheim Church. Just be prepared to catch it if it tries to fly away. Divine protection can be so fickle…

Hot Rocks

Before the discovery of the incredible energy created in nuclear reactions, scientists were forced to attempt to explain various phenomena in the universe in terms of energy sources with which they were familiar. Coal was one source with which every 19th century scientist and housewife had experience, and German physicist Julius Robert Von Mayer used it in a model to try to explain the tremendous energy output of the sun.

(Mayer is best known for sharing the discovery of the first law of thermodynamics, also known as the universal law of the conservation of energy, with James Joule. However, initially, all the credit was given to Joule, and as a result Mayer attempted suicide. Don't ever doubt that scientists take their work *very* seriously!)

Mayer calculated that a burning sphere of coal the size of the sun would only be able to sustain such high temperatures for another 4,600 years—fortunately, a figure that is several billion years short. When he tried to fit burning gases into his calculations, the results were not much better, giving only 5,000 years of sustainable energy.

Apparently fresh out of ideas and possible energy sources, Mayer decided that the heat and light were produced by a host of meteors or space debris; their impacts with the sun creating the necessary energy. Unfortunately for Mayer (who died in 1878), he did not live long enough to realize that what big rocks couldn't do, tiny particles could.

Close Encounters of the Absurd Kind

In 1950, the scientific community made a big mistake. It was in that year that Immanuel Velikovsky published *Worlds in Collision*. The book should have rapidly faded into obscurity, but many scientists banded together against the book in an act of what appeared to be censorship. Velikovsky immediately raised the flag of persecution (comparing himself to Galileo) and drew widespread attention as an alleged victim of a scientific Inquisition.

An enormous amount of time has been wasted writing about Velikovsky's theories in the last five decades—theories which weren't even worth the paper and ink with which they were originally printed. However, in the name of Bad Astronomy, it is necessary to write at least a few more paragraphs.

Velikovsky's basic theories are as follows:

1- The planet Jupiter ejected a comet which passed close to the Earth, causing plagues and catastrophes around 1500 B.C., including showers of burning oil and rocks, locusts, frogs, and other assorted Biblical vermin.

2- This comet threatened us on several more occasions, as did the planet Mars. The results of these close encounters were changes in the Earth's rotation, revolution, and magnetic field, to name a few.

3- The roving comet eventually settled into orbit, lost its flaming tail of doom and became the humble planet Venus.

To support his chronology of catastrophes, Velikovsky did nothing less than rewrite human history to fit his "facts." However, there are some slight glitches in his alleged sequence of events, but so what if there are about a thousand years of recorded history missing from his scheme of things?

What has been presented here is just the tip of the Velikovsky Iceberg of Bad Astronomy, yet it is already evident that he also swam freely in the murky waters of Bad History, Bad Physics, and Bad Geology, as well. It should also be mentioned that Velikovsky was not a professional astronomer; he was a psychiatrist.

Rather than writing volumes of nonsense, his time might have been better served by lying down on his couch and having a very long talk with himself. Perhaps then he would have realized that his ideas were not worth listening to.

"There is no likelihood man can ever tap the power of the atom. The glib supposition of utilizing atomic energy when our coal has run out is a completely unscientific Utopian dream, a childish bug-a-boo."
Robert Millikan, Nobel Laureate in Physics, Chemists' Club, NY,1928

"There is not the slightest indication that nuclear energy will ever be obtainable. It would mean the atom would have to be shattered at will."
Albert Einstein, Nobel Laureate in Physics, 1932

"The energy produced by the atom is a very poor kind of thing. Anyone who expects a source of power from the transformation of these atoms is talking moonshine."
Ernst Rutherford, Nobel Laureate in Chemistry, 1933

Another Nutty Professor?

There's an old expression that says, "If you don't like the weather, wait five minutes." However, some people apparently really can't tolerate Earth's rapidly changing and unpredictable weather patterns. In fact, at least one individual preferred action to waiting—action in the form of blowing up the Moon!

Professor Alexander Abian (1923-1999) was a mathematician at Iowa State University, and he actually suggested that if we destroyed our Moon and changed the Earth's angle of rotation, we would not only rid the world of things like hurricanes and tornadoes, but diseases such as cancer and AIDS, as well. The fact that such actions would also probably rid the world of life (for instance, by way of one good chunk of lunar "shrapnel") didn't appear to be a concern to the old professor. (By the way, Abian proposed to destroy the Moon by drilling "a big hole," filling it with nuclear warheads, "and you detonate it by remote control from Earth.")

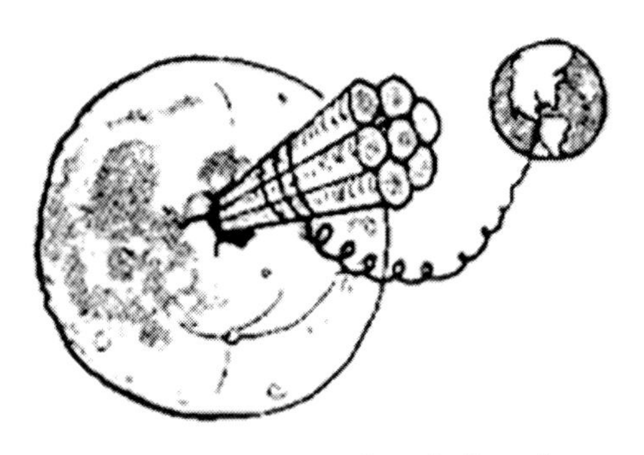
Joseph Gyscek

Another of Abian's celestial remodeling tips involved Venus. He insisted that we must move Venus into an Earth-like orbit to make the planet a "born again Earth." How this was to be accomplished and what impact (literally) this would have on Earth could no doubt fill its own volume of Bad Astronomy.

The final insult to one's sensibilities came not from his theories, but from his response to his detractors (of which there were many). "Those critics who say 'Dismiss Abian's ideas' are very close to those who dismissed Galileo," Abian claimed. (Where have we heard *that* defense before?)

Thankfully, there is little chance that history will ever move Abian into a Galileo-like realm.

"Flight by machines heavier than air is unpractical and insignificant, if not utterly impossible." Simon Newcomb, astronomer, 1902

"I confess that in 1901 I said to my brother Orville that man would not fly for fifty years. Two years later we ourselves made flights. This demonstration of my impotence as a prophet gave me such a shock that ever since I have distrusted myself and avoided all predictions." Wilbur Wright, 1908

One of the Few Times Bigger Isn't Better

The 17th century craze for longer refracting telescopes brought some ridiculous results. Telescopes over 100 feet long were produced, and they were heavy, awkward, and almost impossible to mount on anything sturdy and reliable. Christiaan Huygens eliminated those unwieldy tubes by creating the aerial telescope. The objective lens was placed high atop a long pole, and 123 feet away, the observer held an eyepiece steadied by two wooden legs and attempted to line up the lenses with the help of a thread stretched between them.

As absurd as this sounds, it worked. Of course, it was very difficult finding faint objects and tracking them, but owing to the quality of the 7 ½-inch objective, those with steady hands obtained good images. In 1692, Huygens' brother, Constantine, gave the instrument to the British Royal Society, but they couldn't find a pole long enough upon which to mount it. Almost two decades later, James Pound acquired a secondhand maypole and actually used the shaky aerial telescope to obtain accurate measurements of planets and moons.

Despite the difficulties of handling telescopes like Huygens' 123-footer, he and others made lenses for even longer telescopes. The record appears to belong to Adrien Auzout, who made lenses with focal lengths up to 600 feet! While these mammoth creations were never used, Auzout was hoping that they would enable him to actually see mammoths, or whatever type of animals he believed inhabited the Moon.

Eventually, reflecting telescopes, achromatic, and apochromatic lenses stemmed the focal length insanity, but not before many a man could proudly display his objective and declare, "Mine is bigger than yours!"

Bad Archaeoastronomy

Whatever else Stonehenge has been during its long history, it most certainly was used for astronomical observations (although for what purpose the observations were made may never be certain). However, this aspect of the mysterious stone monoliths was not clearly recognized until 1963 by Gerald Hawkins.

Perhaps this knowledge would have come to light sooner had some people during the Victorian era been more intent on studying the structure than breaking it to pieces. During those years when most of the world was England's playground (and the people its playthings), enterprising individuals rented hammers at the famous ancient site so that visitors could knock off pieces for souvenirs, or just to have a jolly good time bashing away at something they could neither appreciate nor comprehend.

An even worse fate befell a site in Salem, New Hampshire. Once known as Mystery Hill, and now referred to America's Stonehenge, the unusual series of structures and pointed stones was treated as an open quarry for the construction of local towns. This site was not completely destroyed, but one shudders to think of how many other sites have been plundered and plowed over in the name of progress.

And who knows, perhaps our descendants in the distant future will rent lasers to visitors of what was once the state of California, so that they can burn off chunks of Palomar Observatory to show the folks back home.

Friar Fire

The Maya of ancient Mesoamerica developed a unique system of hieroglyphic writing with which they were able to record their complex religious beliefs and practices, as well as their wealth of mathematical and scientific knowledge. One science in which they excelled was astronomy. By the time the Spaniards arrived in the sixteenth century, Mayan texts (or codices) contained tables for predicting eclipses, detailed observations and predictions of Venus' motion, and a 365-day calendar more accurate than its contemporary European counterpart.

However, the Spaniards apparently failed to appreciate even a single word of the collected knowledge of the Maya and strove to eradicate all of its ancient learning. In this cause, one of the Spanish champions of ignorance was Friar Diego de Landa, who never hesitated to use torture in his holy quest against paganism (i.e., any beliefs other than his own). And what he couldn't eradicate by the sword, he destroyed by fire.

In one village in 1562, de Landa gathered all the codices (which were conveniently written on bark) and burned them "since they contained nothing but superstitions and falsehoods of the devil." He recorded that the Maya reacted "most grievously" and with "great pain" at the loss of their books, but they obviously were too stupid to realize it was for their own good.

So effective was this campaign of book burning that of all the codices produced by this ancient civilization, only four survived. (Hopefully, poor Friar de Landa didn't fret the remainder of his life about the ones that got away.) Yet, even from just these four codices it is clear that Mayan astronomers were among the most precise observers and mathematicians in the history of the world. The indiscriminate destruction of their knowledge earns these Spaniards very black marks in the codices of Bad Astronomy.

Friar Diego de Landa did return to Spain to stand trial for his cruelties perpetrated against the native peoples, but not surprisingly, he was not only pardoned, he was made Bishop of Yucatán.

The pious Diego de Landa, who never hesitated to torture Mayans who were hesitant to convert to the loving and compassionate Christian faith.

Is Anybody Home?

Carl Friedrich Gauss (1777-1855) was a gifted German mathematician who began making discoveries in his teens and continued through a long, illustrious career. He was also an astronomer, and was director of the observatory in Gottingen for 48 years. His accomplishments place him firmly in the pantheon of geniuses, yet on at least one occasion, Gauss took a slight flight of fancy.

Assuming that Mars was inhabited, Gauss theorized about how we could contact our Martian neighbors. In 1802, he thought he had found a solution. In the vast tundra region bordering the Arctic Ocean in Siberia, Gauss proposed that huge figures be drawn into the surface. Of course, these figures would have to be large enough for Martian telescopes to resolve, and they needed to be geometric shapes, so there would be no mistake that they were not of natural origin.

As if to get one up on Gauss' "Nazca North" plan, a rather ludicrous suggestion was made in 1874 by Charles Cros. Just in case the Martians weren't looking our way with telescopes, Cros decided to deliver a message to their doorstep, or desert, to be more accurate. By angling the sun's rays through a large, movable lens, Cros wanted to burn his message right into the Martian landscape. Perhaps the idea came to him while frying ants beneath a magnifying glass; a scenario which could have been repeated on a massive, catastrophic scale had it actually been possible to build the contraption and it was pointed the wrong way (e.g., at his house, his friends, the nearby town, etc.). Even if he was successful in sending a burning beam through the far reaches of space, didn't Cros consider that any potential Martians might consider it an act of war?

No one can be faulted for wishing to contact intelligent life beyond our planet; scientists are still trying to this day. Perhaps some alien civilization is even trying to contact us. So if you're ever in the desert and deadly beams of burning light are heading your way, don't forget to smile and wave!

"There is practically no chance communications space satellites will be used to provide better telephone, telegraph, television, or radio service inside the United States." Tunis Craven, FCC Commissioner, 1961

Put Up and Shut Up

All too often people indulge in wild speculations, but few are willing to put their money where their mouths are. In 1952, Godfried Buren not only put forth a ridiculous theory, he put up a lot of cash to back it.

The theory: There is a planet within the sun that is covered in lush vegetation. (Hot pepper plants and burning bushes, perhaps?)

The cash: Twenty-five thousand Deutsche Marks to the first person who could prove that the inside of the sun wasn't a massive greenhouse.

The result: A legal battle that was decided in favor of the astronomical society which offered the proof that disproved Buren's theory.

The tragedy: The only truly sad part in this farce is how much money the lawyers probably made in the process.

Martians Are Easy

Instead of discouraging such bizarre ideas as building bonfires to attract the attention of extraterrestrials, such suggestions were actually encouraged—money being the great motivator.

In France in 1900, a contest was begun for the first person to communicate with a life form from beyond our planet. While fame was no doubt assured for the winner, fortune wouldn't be far behind. The prize was 100,000 francs; more than enough to allow our extraterrestrial neighbors to call collect.

However, there was a catch. Any person communicating with Martians *was not eligible* for the prize money. It wasn't that the contest officials were prejudiced toward inhabitants of the red planet, they simply felt that there wasn't any challenge to contacting Martians. After all, when you're giving away that kind of money you want the contestants to show some effort.

"Transmission of documents via telephone wires is possible in principle, but the apparatus required is so expensive that it will never become a practical proposition."

Dennis Gabor, physicist, author of *Inventing the Future*, 1962

Fine Line

In 1930, Clyde Tombaugh produced photographic evidence for the existence of Pluto. The photos ended a quarter of a century hunt for the tiny planet; a hunt begun by Percival Lowell. In 1905, Lowell had concluded that the orbital perturbations of Uranus and Neptune were caused by an unseen ninth planet, but unfortunately, he didn't live long enough to see it. However, Lowell apparently sought balance in his career by seeing things that weren't really there.

The Lowell Observatory in Flagstaff, Arizona, sits atop Mars Hill, so named for the obsession of its founder. In 1877, Giovanni Schiaparelli reported "canali" or channels on the surface of Mars. Lowell, and most of the rest of the English-speaking world, misinterpreted canali as canals, of the Martian-made variety. Determined to continue Schiaparelli's work, Lowell built his observatory in the clear, thin mountain air (perhaps a little too thin) of the Southwest.

Tirelessly observing Mars for countless hours, he produced numerous, detailed sketches showing an intricate network of fine lines which he believed to be canals—exactly 288 canals bringing water to every corner of the parched planet, and giving life to the cycles of vegetation he also claimed to witness. Cities located at the intersections of these canals further indicated that Martian civilization was very peaceful and advanced, probably more advanced than our own.

In 1895, Lowell published *Mars*, a book detailing his findings. In 1898, he founded a journal devoted to Martian theories, and in 1906, he published another book, *Mars and its Canals*. Together, these works helped to fuel the popular belief of a noble race struggling to survive on a dying planet.

However, some astronomers had little sympathy for our Martian neighbors, and even less for Percival Lowell. Lowell's findings were severely criticized by astronomers like E.E. Barnard of the Lick Observatory, and E.M. Antoniadi, working at Meudon Observatory outside of Paris. They claimed that they had never seen any features remotely like those Lowell was claiming to see. (Interestingly enough, however, a few decades ago it was discovered that both men had made sketches of Mars showing canal-like lines almost identical to Lowell's.)

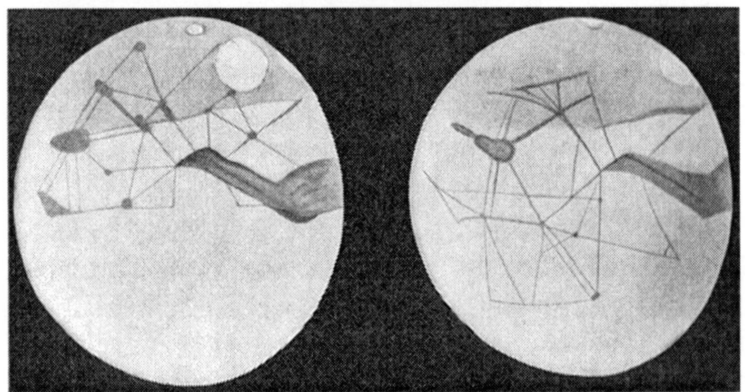

Lowell observing Mars, and his sketches of the alleged canals.

The debate took some absurd turns, with Lowell claiming that his eyesight was "acute," allowing him to detect the fine lines, while Barnard, although known for his superior vision, actually only had "sensitive" eyes incapable of detecting the delicate tracery of the canals.

To Antoniadi's criticism, Lowell countered that the Paris telescope (a 33-inch refractor) was obviously *too big* to obtain accurate views of the red planet. While Lowell had a 24-inch refractor at his observatory, he always stopped it down by six to twelve inches, believing it allowed him to better see the Martian engineering feats.

Surprisingly, the most damaging evidence against Martian life came not from an astronomer, but from octogenarian naturalist Alfred Russel Wallace. It was Wallace's studies of comparative biology that led him to a theory of evolution similar to Darwin's. Wallace, despite his advanced age, turned his considerable talents to Lowell's calculations, and in 1907 uncovered several mistakes; mistakes which turned Lowell's image of an England-like climate with free-flowing water, to the harsh reality of an almost perpetually frozen world with an atmosphere even too thin for the outspoken Lowell to catch his breath.

(However, in all fairness, it must be pointed out that in his analysis, Wallace crossed a line of his own. Extending his conclusions to the rest of the universe, he believed that life couldn't exist anywhere else but on Earth.)

Dismayed, but undaunted by the barrage of criticism, Percival Lowell continued to firmly believe in the Martian civilization until the day he died. For his fervent desire to discover life on other planets he cannot be condemned—it is a desire many of us still harbor. And as for the strange features of Mars he claimed to see, some competent amateur astronomers still occasionally report seeing bizarre shapes. However, the following summary clearly shows where Lowell blatantly leapt over the fine, canal-like lines between science and Bad Astronomy.

What Might Have Been Going on in Lowell's Mind:

1. Now that I've told everyone just exactly what I'm going to find on Mars, I'd better build an observatory and gather lots of evidence to support every one of my claims.

2. My eyes are better than your eyes.
3. Idiots! Don't they realize big telescopes are worthless? I sure hope they never put a big telescope in space. That would be a complete disaster.
4. Everyone must be jealous of my abilities. Yeah, that's it, they're all suffering from Percy-envy.
5. So I was wrong about little details like the harsh climate and lack of water. That doesn't mean I'm not right, does it?

Lunar Roving

On the evening of October 21, 1939, a weather balloon caught fire and fell to earth between Springfield and Strafford, Missouri. The following day a local newspaper, the *Springfield News and Leader*, declared that the Moon had burned and crashed onto the highway.

However, it seems that not everyone in the intellectual mecca of Springfield believed that the Moon had crashed, as they had never heard of this happening in their town before, even though some people had lived there for fifty years.

The more Bad Astronomy changes, the more it stays the same.

"To place a man in a multi-stage rocket and project him into the controlling gravitational field of the moon where the passengers can make scientific observations, perhaps land alive, and then return to earth—all that constitutes a wild dream worthy of Jules Verne. I am bold enough to say that such a man-made voyage will never occur regardless of all future advances." Lee de Forest, inventor of the vacuum tube, 1957

Broken-down Chariots

To point out the errors, inaccuracies, and exaggerations in Erich Von Daniken's books—*Chariots of the Gods, Gods From Outer Space,* and *Gold of the Gods*—would in itself, require an entire book. In fact, it has already been done at least once (*Crash Go The Chariots* by Clifford Wilson). Even concentrating merely on the

astronomical errors would be a lengthy task, so only two brief examples will be given.

1. In *Chariots*, in order to bolster support for his idea that the pyramids were built by visitors from space, Von Daniken poses the question, "Is it really a coincidence that the height of the pyramid of Cheops multiplied by a thousand million—98,000,000 miles—corresponds approximately to the distance between the earth and the sun?"

As the actual distance is about 93 million miles, giving a hefty 5 million mile error, it appears that our visitors weren't very good at simple calculations, but then neither was Von Daniken. The height of the pyramid is 481.4 feet, which in his arbitrary equation would actually yield a result of 91.17 million miles. (Is it really a coincidence that both Von Daniken and the space men have poor math skills?)

2. Early on in *Chariots*, Von Daniken refers to "a cave drawing (that) reproduces the exact position of the stars as they actually were 10,000 years ago. Venus and earth are joined by lines."

Later in the book, he throws his hat in Velikovsky's ring and claims that the data from Mariner II "confirms Velikovsky's theory"—the theory being that "a giant comet crashed into Mars and that Venus had been formed as a result of this collision." This is important to Von Daniken, because he believes this would support his "thesis that a group of Martian giants perhaps escaped to earth to found the new culture of homo sapiens by breeding with the semi-intelligent beings living there."

Now, according to Velikovsky, Venus was formed from Jupiter, but paternity aside, the important number is his claim that Venus was formed about 3,500 years ago. So how can Von Daniken, in the space of one book, agree with the theory that Venus is only 3,500 years old, and also refer to a 10,000 year old drawing of Venus? How clever these spacemen were to draw a planet that wasn't going to form for thousands of years.

The first line of the introduction to *Chariots* states, "It took courage to write this book, and it will take courage to read it." Courage, and a strong stomach. In addition to his imaginative use of facts, Von Daniken has sucked the life out of humanity's creative intelligence, reducing mankind to cowering lab animals who couldn't discover the noses on their faces without the help of aliens.

The theory that earth has been visited (and is still being visited) by beings from other planets is, at the very least, a compelling concept. However, if perverting astronomy, archaeology, and human nature is necessary to prove this theory, it would be better to believe that we are all alone in the universe.

"Space travel is utter bilge." Dr. Richard van der Reit Wooley, space advisor to the UK government, 1956

"Space travel is bunk."
Sir Harold Spencer Jones, Astronomer Royal of the UK, 1957
(Two weeks after this statement, Sputnik was launched.)

Not Worth the Ink

From the day Neil Armstrong and Buzz Aldrin walked on the surface of the moon in July of 1969, there have been those who claimed it was all faked and staged.

That's just too stupid even for this book to waste the time and ink writing about such idiotic claims.

Let's Face It

Viking 1 provided thousands of pictures of the surface of Mars, but one in particular created an alien civilization craze unparalleled since the days of Percival Lowell. However, the issue has gone far beyond mere canals—we are now dealing with mile-long faces, towering pyramids, and an entire ancient Martian city, or so some people claim.

The photograph that started the stir was taken in June of 1976, and showed a region (41' N latitude, 9.5' W longitude) that contained some odd features, including one resembling a face. The picture was examined and after concluding that it was an interesting, but natural, geological formation, it was filed away. The actual stir didn't begin until 1980, when Vincent DiPietro and Gregory Molenaar resurrected the photo, computer-enhanced it, and concluded that the face did not seem to have been made by "totally natural forces." They presented these findings at the American Astronomical Society meeting in June of 1981.

Things began to escalate, and soon there was talk of a city with pyramids, forts, amphitheaters, and water tanks. The mouth of the

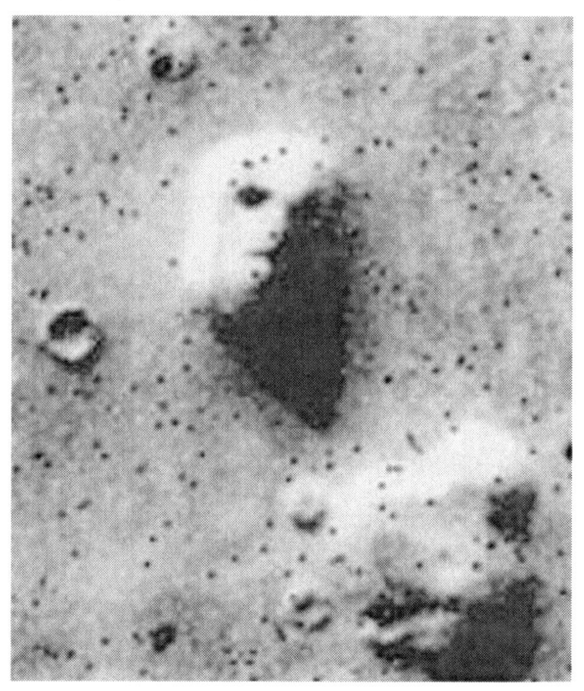

The controversial Face on Mars.

face apparently is not only aligned with the heart of the "city," but the rising of the sun at the summer solstice as well—at least where it would have risen half a million years ago, when this culture supposedly thrived in a more hospitable climate. Assumptions were heaped upon assumptions and any semblance of reason and scientific method was lost in a circus-like arena of speculation.

The entire Face on Mars mess was supposed to be resolved when the Mars Observer was scheduled to enter orbit in August of 1993. The pro-facers claimed that conclusive evidence of the city would finally be provided, while the other side hoped the issue would once and for all be defaced. Unfortunately, it is suspected that as the fuel tanks of the Observer were being pressurized in preparation for entering orbit, there was a leak that caused the craft to spin out of control. Of course, this was only NASA's explanation—the pro-facers had some very different ideas. Yet even they couldn't agree on the same scenario.

To briefly summarize:

1. The Mars Observer blew up, because our government intentionally blew it up to hide the truth of the alien civilization. (Among the many questions here is why not just eliminate the budget for the project in the first place, rather than spending an enormous amount of time and money to build, launch, and monitor something you intended to blow up?)

2. The Mars Observer is in its proper orbit and is currently sending secret information back to earth.

3. The Mars Observer was destroyed, only we didn't do it. The craft, like so many before it that were never heard from again, was knocked out of commission by nervous Martians who were afraid we would see too much. (So why have they let us see this much?)

No rational individual could seriously believe or hope that we are alone in this vast universe. While the discovery of an ancient civilization on Mars would be the greatest discovery of all time, jumping to irrational conclusions will never help the credibility (or funding) of research into this possibility.

The controversy should have been solved in April of 1998, when the Mars Global Surveyor took a much more detailed image of the Martian enigma. The higher resolution image clearly showed that the face was a natural mesa or butte, similar to those throughout America's western landscapes. Critics, however, said the photo was taken on a cloudy day and obscured the Martian-made details.

Once again in 2001, NASA obliged the conspiracy-crazed public and took another image. This photo had a resolution of 1.56 meters per pixel, which was a vast improvement over the 43 meters per pixel in the original 1976 Viking image. To the dismay of the pro-face community, the picture revealed a weathered lump of rock and dirt, with no signs of alien intervention. No pyramids, no water tanks, nothing but naturally formed features.

However, let's face the real facts. As history has repeatedly shown, no amount of evidence will ever convince the adamant skeptic, as no amount of debunking could ever daunt the firm believer.

If there is a lost civilization on Mars, it would come as no surprise to discover that its demise was the result of one group of stubborn Martians fighting with an equally intractable group of obstinate Martians holding opposing views. But hey, isn't that what life is all about?

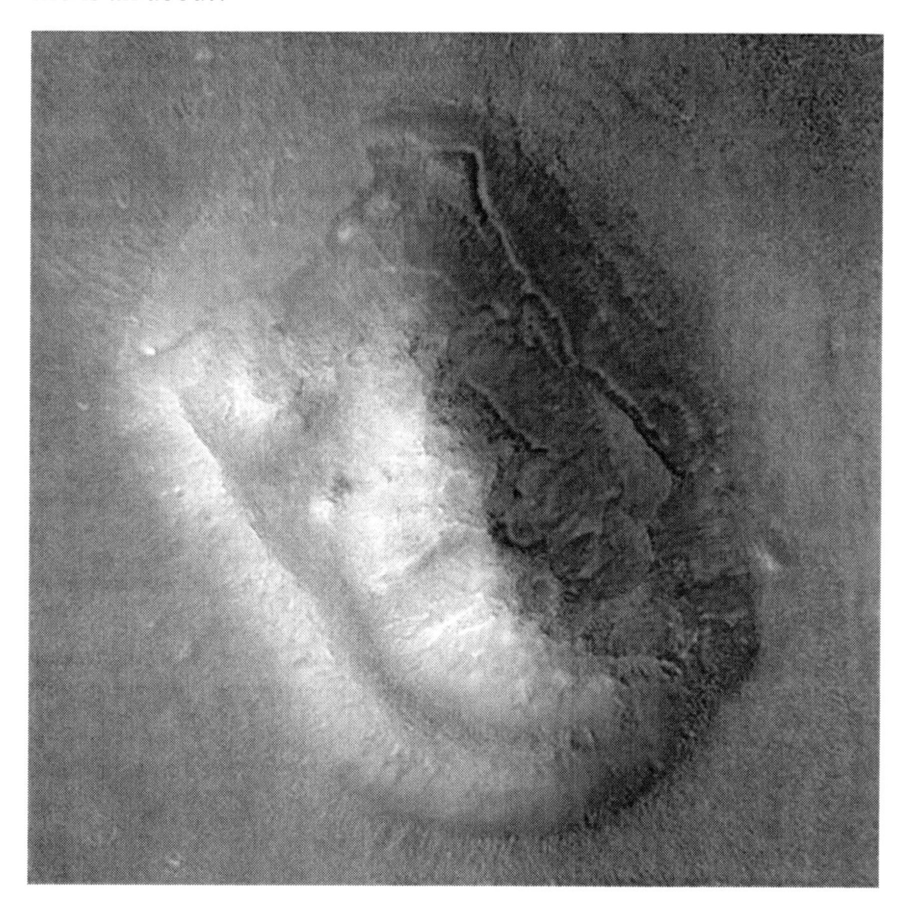

The high resolution image of the Face on Mars, which shows no face on Mars.

"There is no reason anyone would want a computer in their home."

Ken Olson, president, chairman, and founder of Digital Equipment Corp., 1977

Star War?

**HERE MEN FROM THE PLANET EARTH
FIRST SET FOOT UPON THE MOON
JULY 1969 A.D.
WE CAME IN PEACE FOR ALL MANKIND**

These were the words on a plaque that was on the leg of the Apollo 11 lunar lander. A noble sentiment, indeed, and it's a pity that this notion of peace was not shared by everyone who ventured into space.

In the 1950s, competition with the Soviets had given birth to the space race. From the instant Sputnik was launched until the moment Neil Armstrong stepped onto the Moon's surface, the desire to be the first nation to claim the glory of the successive achievements was intense.

However, on July 17, 1975, when the joint Apollo-Soyuz mission saw astronaut Tom Stafford and cosmonaut Alexei Leonov shake hands in space and smile for the cameras, it was supposed to be the culmination of five years of cooperation and planning, and usher in a new age of mutual trust. (Sounded good on paper, even if both countries still had bundles of nuclear warheads pointed at one another.)

Apparently, though, reality was a bit different. The veil of secrecy that had shrouded many of the missions in the Soviet space program finally began to lift in the 1990s, and a startling discovery was made about their space station *Salyut 3,* which was in orbit from June, 1974, to January, 1975. Like a celestial battle ship, *Salyut 3* packed some serious hardware, in the form of a Nudelman-Rikhter 23mm rapid-fire cannon! (Or possibly a 30mm version, but that's not the point.) This machine gun was affixed to the hull of the station "for defense against U.S. space-based inspectors/interceptors."

Were they actually afraid that American space pirates were going to board their station and then blast off with a treasure of stolen technology?

Can you just imagine:

"Houston, this is Blackbeard One. We have scuttled the Soviet station. Arrrgh. Over."

"Roger that, Blackbeard One. Crack open a flagon of Tang to celebrate a job well done."

What the Soviets should have feared was actually firing a machine gun that's bolted to the hull of your space station. There's the small matter of Newton's Third Law of Motion—for every action there's an equal and opposite reaction—and the recoil of a 23mm cannon could wreak havoc on one's orbital trajectory. There were claims that special anti-recoil engines had been installed to compensate and that there had been several successful test firings, but even a slight error could have dire consequences in more than one direction:

1. Leaving orbit and heading into outer space to face a freezing death.
2. Leaving orbit and entering the earth's atmosphere to face a burning death.

Orbital death spirals aside, there was still the matter of the practical aspects of the machine gun. It was not mounted on a turret or other rotating device, so the only way to aim the thing was to point the entire space station toward the evil American invaders while looking through a periscope trying to line up the target.

As cosmonauts also carried pistols, perhaps it would have been easier to simply knock out a window and start shooting!

Today, the International Space Station is a symbol of true cooperation among nations, but let us not be fooled by this one shining example. Space is bristling with military hardware, and while it is unlikely that satellites are toting machine guns, they probably have technology far more deadly.

So perhaps in retrospect, we should applaud the former Soviet Union for its foresight in arming its space vehicles. As mankind reaches farther into space, we are truly shooting for the stars, in more ways than one...

Deep Disturbance

One of NASA's boldest plans and greatest successes came to fruition on July 4, 2005, when a copper-tipped probe slammed into the surface of Comet Tempel 1. The mission, known as Deep Impact, had to send the spacecraft on a precise intercept course to catch the moving comet and strike it with the probe—all at a distance of 83 million miles from Earth. Talk about threading a needle on an interplanetary scale!

In addition to providing some spectacular fireworks for astronomers and the Deep Impact scientists on the Fourth of July, the team has been able to analyze the data collected from the cometary explosion, which has resulted in some important discoveries. They found both water ice and organic material (crucial ingredients for life as we know it), and were also able to determine from the composition of the comet that it most likely originated in the region of Uranus and Neptune.

Remarkable stuff! So who could possibly object to this triumph of science? And why would anyone want to file a lawsuit against NASA because of the project?

Try a Russian astrologer, one Marina Bai, who claimed that punching a crater into the surface of the Tempel 1 comet would do nothing less than "disrupt the natural balance of the universe." And for this great offense, Bai wanted to be compensated to the tune of $300 million!

The astrologer's lawyer stated, "My client believes that the NASA project infringes upon her spiritual and life values as well as the natural life of the cosmos and would disrupt the natural balance of forces in the universe." Remarkably, a Russian court approved the first step in the angry astrologer's quest for more money than the entire Deep Impact project cost.

The case commenced in Moscow in September of 2005, but was dismissed when physicist Vladimir Fortov asserted that the collision didn't have the slightest effect on the Earth, and that "the change to the orbit of the comet after the collision was only about 10cm"—or about $300 million short of creating a deep disturbance.

> Bright points in the sky or a blow on the head will equally cause one to see stars.
> Percival Lowell, *Mars*, 1895

It Doesn't Take a Rocket Scientist…

Against stupidity the very gods
Themselves contend in vain.
-Schiller

A ten-year-old child would know better than to load old operating software into a new computer system. Unfortunately for the European Space Agency, they did not have any ten-year-olds on their review committees for the Ariane-5 rocket. If they had, perhaps the maiden flight would not have ended in disaster.

Like the Ariane-4, its reliable predecessor, Ariane-5 was the product of Arianespace, a commercial company established in 1980. After a decade of research and development and expenditures exceeding $8 billion, the new rocket was supposed to revolutionize the satellite industry by doing more while costing less; truly a rare feat in the modern world.

What massive catastrophe brought down the mighty rocket? Was it sabotage? Was it severe weather conditions? Was it a freak accident that never could have been prevented?

Or, could it possibly be that they loaded the old software from Ariane-4 into Ariane-5, and then didn't bother to test it on all the new systems? Could anyone possibly make that kind of mistake with $8 billion dollars and lifetimes of work on the line? Incredibly, yes.

At 12:33 GMT, on June 4, 1996, the unmanned Ariane-5 was launched from the European Space Agency's facility in Kourou, French Guiana. The first 37 seconds after ignition looked flawless. Suddenly, at an altitude of 3,700 meters, the rocket veered sharply. The stress began to tear it apart and to protect lives on the ground, the self-destruct system blew Ariane-5 to bits.

To make matters worse, Ariane-5 was carrying the *Cluster* project—four satellites designed to study the sun's effects on the Earth's climate. The project had employed over 500 scientists working over a span of ten years, at a cost of $500 million.

Nothing had been insured.

Technically, the disaster occurred because two computers (Inertial Reference Systems) used to guide the rocket shut themselves down due to a software error: the computers' inability to convert a 64-bit floating point to a 16-bit signed integer value. In the words of the official report of the Inquiry Board, the old Ariane-4 software that caused the error actually "serves no purpose" on the new Ariane-5 after launch, and never should have been active.

In simple terms, the wrong software operated at the wrong time and sent the wrong information. (And in even simpler terms for those who are truly technically challenged: software bad, make big stupid mistake, make rocket go BOOM!)

Among the Inquiry Board's brilliant recommendations were to "Review all flight software" in the future, turn off anything that isn't needed, and test all systems with "realistic input data." Duh. One shudders to think what the original testing protocol required. Perhaps such crucial things as, "Make sure there are no pieces left over after you put the rocket together" or "Never test today what you can blow up tomorrow."

The conquest of space is truly one of mankind's greatest endeavors, and no undertaking of this magnitude can be expected to be without disasters. However, in the painful light of colossal blunders such as Ariane-5 and the Hubble Space Telescope, perhaps a simple motto should be sown onto the lab coats of all the project scientists: "Try it before you fly it!"

Have You Ever Been Plutoed?

For many years, the search for a ninth planet yielded nothing, until Clyde Tombaugh finally found it in 1930. Named Pluto, generations of students grew up reciting the nine planets in our solar system. Unfortunately, even planetary families have an occasional falling out, and this runt of the litter was destined to be kicked to the astronomical curb.

On August 24, 2006, the International Astronomical Union created a new definition of a planet, and due to its diminutive size, Pluto didn't make the grade. This created quite a stir amongst scientists and the general public, who didn't take kindly to having one of their planets taken away. However, the solar system's loss turned out to be the English language's gain.

In 2006, the word of the year selected by the American Dialect Society was "plutoed"—meaning to "demote or devalue someone or something." It has become a popular phrase, especially amongst students, that when you have been taken down a peg you have officially "been plutoed." While the former ninth planet may not be as big as the rest of its siblings, no other planet from Mercury to Neptune can claim to be so trendy!

Robert Goddard built and launched the first liquid-fueled rocket on March 16, 1926. In his writings, he suggested the possibility of sending a rocket to the Moon. How did the press treat this brilliant "Father of Rocketry"? With ridicule, of course:

"That Professor Goddard, with his 'chair' in Clark College and the countenancing of the Smithsonian Institution, does not know the relation of action to reaction, and of the need to have something better than a vacuum against which to react - to say that would be absurd. Of course he only seems to lack the knowledge ladled out daily in high schools."

New York Times, January 13, 1920

The *New York Times* did not issue a correction until Apollo 11 was on the way to the Moon!

"Further investigation and experimentation have confirmed the findings of Isaac Newton in the 17th Century and it is now definitely established that a rocket can function in a vacuum as well as in an atmosphere. The *Times* regrets the error." *NYT*, July 17, 1969

Hubble, Hubble, Toil and Trouble

(A tragedy of Shakespearean proportions.)

Background: The optics company Perkin-Elmer ground the mirror of the Hubble Space Telescope, but had incorrectly assembled a device called a null corrector that was used to measure the mirror's critical shape. Two properly functioning null correctors accurately indicated the mirror was not shaped right, but they ignored those results. Only after the telescope was launched in 1990 did the error literally come to light as the world gasped at Hubble's blurry images. A service mission of the shuttle *Endeavor* was launched in 1993 to repair the defective optics.

ACT 1
Scene I- A deserted alley behind Perkin-Elmer, Danbury, Connecticut, early 1980s.

Thunder and lightning. Enter three optical engineers.

First Engineer When shall we three meet again
In thunder, lightning, or the
appropriations subcommittee inquiry?

Second Engineer When the hurlyburly's done,
When the mirror's lost and won.

Third Engineer A decade at least then, til the
Endeavor's had its run.

The three join hands and circle a dumpster.

All Fair is foul, and foul is fair;
If they don't test it,
They must not care.

In Conclusion:
The Mother of All Bad Space Travel
(or Lack Thereof)

I have been cheated. I have been robbed, sold a bill of goods, been led down the primrose path, only to have my hopes and dreams ground down under the heel of ignorance.

When I was a kid, I had astronomical charts on my bedroom walls, built models of rockets, and couldn't get enough of all things

related to outer space. I was riveted to the television the night Neil Armstrong's foot first touched the lunar surface.

I couldn't believe how lucky I was to be growing up during mankind's greatest age of exploration and discovery, during a time when courage, determination, innovation, and technology brought the world together to witness our greatest achievement as Homo sapiens. And that was just the beginning, as we were all to pull together to establish bases on the Moon, walk upon the red soil of Mars, and extend our reach beyond our wildest dreams. What wonders I would see if my lifetime!

Then they cancelled Apollo 18, 19, and 20. Instead, we had Skylab and space stations and shuttle launches. Lots and lots of shuttle launches, but we never left Earth's doorstep again. I grew up, lost hope, became angry and bitter that the dreams of youth had been shattered.

Finally in 2004, there was a glimmer of hope when President Bush announced that we were returning to the Moon and then heading for Mars. He summed up the space program perfectly when he said, "The desire to explore and understand is part of our character and that quest has brought tangible benefits that improve our lives in countless ways."

Yes, yes, yes! Space exploration had brought out the best in us, had fostered cooperation, had led to so many improvements and inventions that we all use every day. Finally, the people in charge realized the full value of setting lofty goals that bridge the gaps of human frailties and lead to better things we can't even as yet imagine.

We were further assured by Presidential candidate Obama in a speech to space industry workers in Florida in August of 2008, when he stated that, "One of the areas we are in danger of losing our competitive edge is in science and technology, and nothing symbolizes that more than our space program. You know, I've written about this in my book. I still remember sitting on my grandfather's shoulders as some of the astronauts were brought in from their capsules that landed in the Pacific. I could barely see them, but I remember waving the American flag, and my grandfather explained that this is what America is all about. We can do anything we put our minds to. When I was growing up, NASA inspired the world with achievements we are still proud of.

"So let me be clear. We can not cede our leadership in space. That's why I'm going to close the gap, ensure that our space program doesn't suffer when the shuttle goes out of service. We are going to continue to support NASA funding, speed the development of the shuttle's successor, by making sure all of those who work in the space industry in Florida do not lose their jobs when the shuttle is retired, because we can't afford to lose their expertise."

Then President Obama dropped the ax on projects vital to our return to the Moon. As the shuttle missions drew to a close, tens of thousands of aerospace workers lost their jobs, taking their vital expertise anywhere they could find work.

In a perfect summation of the ignorance that kills science, Representative Nancy Pelosi said, "If you are asking me personally, I have not been a big fan of manned expeditions to outer space in terms of safety and cost."

Well guess what, Nancy, I'm not a big fan of yours, either. The fact is, that for every dollar invested in the space program, we have realized seven dollars in economic payback. A seven-to-one return on investment is not bad in any economy! And maybe Nancy and all the others detractors should try living without all the advances in computer technology, telecommunications, medical technology, and industrial innovations brought about by space programs.

In May of 2010, NASA's Human Spaceflight Plan was reviewed by the U.S. House Science and Technology Committee. Former astronauts Neil Armstrong, James Lovell, and Eugene Cernan spoke before the committee of the importance of manned exploration. Armstrong's words encapsulated the frustration and incredulity surrounding the same lack of vision that cancelled the Apollo missions, and threatens to keep us from going any further than Earth orbit.

"Some question why Americans should return to the Moon. 'After all,' they say, 'we have already been there.' I find that mystifying. It would be as if 16th century monarchs proclaimed that 'We need not go to the New World, we have already been there.' Or as if President Thomas Jefferson announced in 1803 that Americans 'need not go west of the Mississippi, the Lewis and Clark Expedition has already been there.'

"Americans have visited and examined six locations on Luna, varying in size from a suburban lot to a small township. That leaves more than 14 million square miles yet to explore."

Not to mention the rest of the solar system.

I have no doubt that history will judge us harshly. With more guts than technology, mankind walked on the Moon. Yet, even as we were crossing the threshold to new worlds, we stopped. We pulled back and changed our priorities. Take a look at the world now, how has that worked for us?

I'm not saying that by investing our time, money, and brainpower in the space program we will cure the world's ills and we will all live peacefully and happily together forever. What I am saying, is that one of our best hopes for a better world is in advances in technology, mutual cooperation, and just maybe focusing our attention on things beyond our own petty squabbles.

I have been cheated out of decades of spectacular discoveries. I have been robbed of the thrill and wonders of space exploration. I have been without the inspiration and admiration of heroes and heroines who never boldly went where no one had gone before. I am sick and I am tired of waiting for the government and general public to awaken from their intellectual comas and reenergize their spirits, wills, and imaginations.

I want to stand on the NASA causeway on some clear, crisp dawn, watch the blinding flash of rocket engines roaring to life, and be buffeted by the crackling sound waves as I experience the launch of a mighty manned vehicle bound for the Moon. Then I want to do it again as we head for Mars.

Poet Robert Browning wrote, "A man's reach should exceed his grasp."

It's time again for us to reach for the stars.

> Don't tell me that man doesn't belong out there. Man belongs wherever he wants to go -- and he'll do plenty well when he gets there.
>
> What is the most important thing a man needs to build? The will to do it.
>
> Wernher Von Braun

The author with astronaut James Lovell
at the Kennedy Space Center in 2009.

Scientists, Heredity, DNA, Firearms, and Everything Else that didn't Fit into Previous Categories

Spontaneous Generation

For thousands of years, people believed in spontaneous generation—the supposed phenomena of living organisms arising out of inanimate matter. For example, it was assumed that maggots were created by rotting meat.

While one can excuse, to some extent, this belief where tiny or microscopic organisms are involved, it seems beyond comprehension that people believed that mice were also generated spontaneously. (As if they needed any help breeding!) As "proof" of this, in the 1600s one brilliant observer of nature described a simple experiment: Take an open jar, add wheat husks and sweaty underwear, and voilá, in three weeks you would have mice! (I don't even want to think about how he came upon that tasty recipe…)

Finally in 1668, an Italian doctor, Francesco Redi, attempted to disprove spontaneous generation, and prove his belief that maggots in decaying flesh were actually the result of flies laying their eggs. In a refreshingly modern experiment that actually used control samples, Redi placed pieces of meat in both open and closed containers, as well as covering some pieces with gauze. Lo and behold, only the exposed pieces of meat became infested with maggots.

The success of this experiment should have placed the idea of spontaneous generation on the short road to extinction, but unfortunately, the theory remained in high gear on the superhighway of Bad Science. Remarkably, even Redi still believed spontaneous generation was the explanation under some circumstances.

One would have hoped that when the microscope hit the scientific community, it would have revealed the tiny secrets of life—that microorganisms and bacteria were behind the mysterious proliferation of life in puddles, old soup, and last week's dinner no one bothered to clean up. However, the reaction was quite the opposite. Rather than disprove spontaneous generation, the discovery of this microscopic world only reinforced the erroneous beliefs, as these "animalcules" were also thought to be created out of inanimate matter.

Once again, someone came up with a simple experiment to prove the theory: Take an open container, add water and hay, and in a few days you could observe an entire little microscopic world of activity. (At least they left out the sweaty underwear this time!)

It wasn't until 1859 that spontaneous generation was finally put out to pasture—or Pasteur, to be more accurate. Thanks to that beacon of truth and knowledge, Louis Pasteur, the theory was convincingly disproved. By boiling meat broth to kill the bacteria in specially designed flasks with down-turned, S-shaped necks, he showed that nothing would grow if airborne bacteria couldn't reach the broth—not to mention also demonstrating the existence of airborne bacteria!

Even in the vast realm of Bad Science, two thousand years is a long time for a blatantly wrong theory such as spontaneous generation to exist. Unfortunately, until the day arrives that spontaneous intelligence arises, such things will continue to be as foul and commonplace as unwashed underwear.

"Louis Pasteur's theory of germs is ridiculous fiction."

Pierre Pachet, British surgeon and Professor of Physiology, 1872

Pasteur

This entire book could probably be filled with stories of the Bad Scientists who plagued Louis Pasteur during his long and glorious career. They fought him on his ideas about fermentation, spontaneous generation, the germ theory of disease, his rabies vaccine, his anthrax vaccine, and just about everything else they couldn't wrap their ignorant little brains around.

Pasteur was born in 1822 in France. He initially planned to be an art teacher, but it's fortunate for humanity that chemistry caught his attention. However, since Pasteur was a "mere" chemist, medical doctors didn't want to hear about his work on human and animal diseases, because he was obviously not qualified. Ironically, some of these same doctors were still spreading childbed fever and other infectious diseases because they didn't wash their hands.

Many prominent men in the medical world called Pasteur's research into microbes nothing short of "madness," even though some doctors absolutely refused to even look through a microscope and see the evidence for themselves. (Just like Galileo's detractors who refused to look through his telescope.)

Also, the more he delved into microbiology, the more he was scorned by other chemists. One of Pasteur's most outspoken critics was German chemist Baron Justus von Liebig. Pasteur traveled all the way to Munich to personally address Liebig's objections to his findings, but the stubborn German brushed off his French rival and refused to discuss the matter. (Liebig claimed he was ill, perhaps a touch of acute badscientisis?)

A lot of other Germans didn't like Pasteur, either, but the feeling was mutual (a 19th century war didn't warm relations between the two countries, and it certainly wouldn't get any better in the 20th century). Then there were the naturalists who didn't care for Pasteur's ideas, a number of Englishmen who were quite put off by his theories, and then several veterinarians added their animosity into the mix just for good measure.

Unfortunately for Pasteur, he was not just introducing a few theories, he was helping to create entirely new fields of science. At least most Frenchmen liked him—except for those dirty-handed

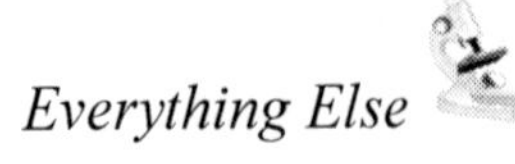

doctors—possibly because Pasteur had helped save the wine industry…

In any event, rather than dirty my hands with the long and unpleasant saga of the multitudes who fought against Pasteur, let me simply applaud the brilliant and tenacious Frenchman who fought for truth, humanity, and damn Good Science.

Note: Just to prove how good Louis Pasteur was, he had a starship named after him on *Star Trek: The Next Generation*. If that isn't the ultimate revenge against his critics, I don't know what is.

An 1885 painting of Louis Pasteur in his lab.

Tanks for the Memories

"Caterpillar landships are idiotic and useless. Those officers and men are wasting their time and are not pulling their proper weight in the war."

Fourth Lord of the British Admiralty, 1915

> "The idea that cavalry will be replaced by these iron coaches is absurd. It is little short of treasonous."
>
> Aide-de-camp to Field Marshal Haig during tank demonstration, 1916

Shooting Your Mouth Off

From the humble prehistoric club to nuclear weapons, if there's one area where mankind has always demonstrated proficiency, it's in ever-efficient ways to kill one another. The development of firearms arguably has had the biggest impact on the history of warfare, so it might prove valuable to present a brief chronology of advancements.

In the 1400s, the Matchlock musket made its debut. The mechanism included a length of burning wick that would fall onto the exposed flash pan (which contained gun powder) and with any luck, the subsequent ignition would fire your gun and kill your enemy. In reality, the wick was often difficult to keep lit—and impossible in damp and rainy weather. The weapon was inaccurate and unreliable, and was more of a means of intimidation. Think of it this way—if the Matchlock muskets were any good, why did the Three Musketeers always use their swords?

A great improvement was made in 1517, when the burning wick was replaced by a piece of flint and a wheel lock mechanism, that when turned against the flint produced a spark that hopefully ignited your powder. Unfortunately for the common soldier, the Wheel Lock was a more complicated and expensive weapon to produce—about twice the cost of the lower quality Matchlock—so it became a weapon of the wealthy, often bearing fancy engravings and inlays.

In 1570, the Snaphaunce appeared, which was basically an early form of the more famous and successful Flintlock, which was invented in 1612. The Flintlock was more reliable and accurate, less prone to bad weather, and could be produced in a cost-effective manner. After all, just because you want to kill a lot of people, it doesn't mean that you shouldn't keep an eye on your budget. Perhaps the most well-known Flintlock is the Brown Bess, the legendary weapon of the Redcoats during the American Revolution (for all the good it did them!).

However, for all its almost 400 years of improvements, firearms still had major flaws. But in 1805, a brilliant invention set the gun in a new direction—the percussion cap. Gone were the clumsy pieces of flint and exposed flash pans. Instead, there was a small metallic cap, shaped something like a top hat, which contained fulminate of mercury. When struck by the hammer (that's the percussion part) the cap produced a spark that you could count on in just about any kind of weather.

(Note: This also led to another great invention, the child's "cap" gun.)

As revolutionary as the percussion cap was, however, it should not have lulled people into a false sense of security that mankind would never devise any more efficient means of killing. And it certainly shouldn't have convinced firearms experts that the absolute zenith of gun technology had been reached. But apparently, it did...

From the *American Shooters Manual* of 1827:

"The lock has undergone a variety of alterations until it would seem that but little room is left for further improvement. Nor do I consider any further amendment necessary as every useful purpose appears to be completely answered by the common lock adapted to the percussion primer."

Well, there we have it, end of story! In 1827, every problem with firearms had been "completely answered."

But wait, someone must have forgotten to tell Smith & Wesson in 1854 when they invented the self-contained metallic cartridge that combined the primer, powder, and projectile into one unit that was to be utilized so effectively in the Henry, Sharps, and Spencer rifles during the Civil War. And apparently, Mr. Gatling didn't get the memo, either, before he invented the gun that brought the rate of fire up to 200 rounds per minute (under ideal circumstances, the average soldier with a percussion cap musket managed about 2 to 3 rounds per minute).

Then there were the machine guns of World War I, Thompson's nifty little submachine gun that put 700 rounds per minute into the hands of gangsters and bootleggers in the 1920s, and the modern-day Uzi with small, lightweight models capable of 1,250 rounds per minute. (But wait, why not have one in each hand to boost that to a very persuasive 2,500 9mm rounds of death every sixty seconds?) And while it may not be the fastest gun, special mention must be made for the sweet sound of the chain guns mounted on the Apache Longbow helicopters that chew up targets with 625 rounds per minute of hefty 30mm ammunition.

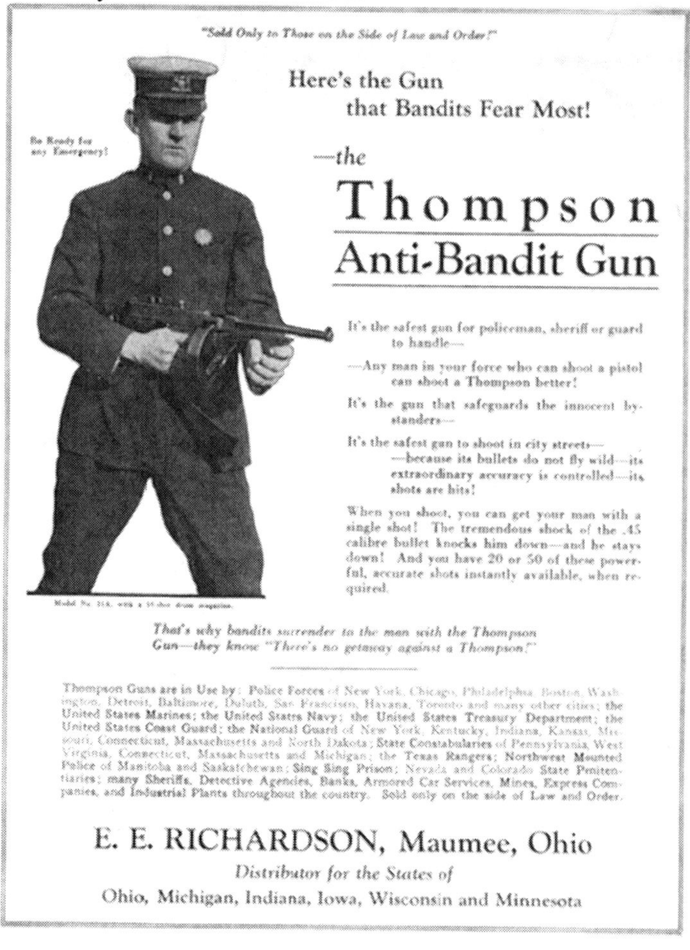

The only problem was that the bandits had them, too!

However, even all of these weapons pale in comparison to the Metal Storm, the brainchild of a former grocer from Australia. (Don't scoff at the fact that he's a former grocer—just remember what a couple of bicycle mechanics did for aviation.) By replacing relatively slow moving parts with "electronic ballistics," the Metal

Storm may be capable of firing a staggering *one million* rounds per minute. Yes, that's a one followed by six zeroes. It's a number that has caught the attention of the Pentagon, and no doubt has munitions manufacturers salivating.

Hopefully, this brief history of firearms has truly demonstrated that no limits should ever be set on mankind's ingenuity (or thirst for blood). And no matter what your field of alleged expertise, don't ever shoot your mouth off!

"Atomic energy might be as good as our present-day explosives, but it is unlikely to produce anything very much more dangerous."

Winston Churchill, British Prime Minister, 1939

"That is the biggest fool thing we have ever done ... The bomb will never go off, and I speak as an expert in explosives."

William D. Leahy, U.S. Admiral, advisor to President Truman on the research into an atomic bomb, 1944

William Charlton

In 1702, William Charlton sent a specimen of a previously unknown butterfly to noted entomologist James Petiver in London. It was a good thing Charlton sent it when he did, as he died shortly after. However, as an avid butterfly collector, Charlton must have found great comfort on his deathbed in the fact that his new specimen would guarantee his name in the annals of entomology.

Petiver was delighted, and recorded his analysis of the new specimen: "It exactly resembles our English Brimstone Butterfly (R. Rhamni), were it not for those black spots and apparent blue moons on the lower wings. This is the only one I have seen."

There was a very good reason why it was the only one Petiver had ever seen, because Charlton had taken an ordinary Brimstone Butterfly and painted the spots on it.

Unfortunately, the fraud did not end there. In 1763, Carl Linnaeus—physician, botanist, zoologist, and "The Father of Taxonomy"—also examined the rare butterfly. He also failed to recognize it as a fake, and decided it was truly a new species. He called it *Papilio ecclipsis* and added it to the 12th edition (1767) of his landmark work *Systema Naturae*. It wasn't until 1793 that a Danish entomologist examined the butterfly at the British Museum and realized the spots on the wings had been painted, which finally put a stop to further efforts to find another like it.

It also put a stop to the butterfly being in the museum's collection, as when news of the counterfeit species reached Dr. Gray, the man in charge of the collection, he threw it to the floor and "indignantly stamped the specimen to pieces." (Let that be a lesson to you about pissing off museum curators!)

While it is sad that Charlton's desire to contribute to his hobby led to learned men wasting time on a fraud, it is also inexcusable in the hallowed museum halls of Good Science. Today, Charlton's Brimstone Butterfly is a symbol of vanity over truth, and perhaps Mr. Charlton is now encountering brimstone of another type, in that place where all bad scientists are condemned to spend eternity…

Perpetual Fraud

One of the Holy Grails of science has been to build a perpetual motion machine—essentially something that when set into motion continues to move without any power source. Despite the fact that the very concept violates one or two laws of thermodynamics, people still seek to make the impossible possible. Then there are those who seek to make the impossible profitable.

In Philadelphia in 1812, Charles Redheffer began charging the hefty sum of $5 per man and $1 per woman to view his perpetual motion machine. Considering that some workers only made *$5 per week*, Redheffer was doing pretty well for himself with his machine. But then he got greedy (or even greedier, depending on your point of view).

Redheffer wanted to build a bigger and better perpetual motion machine, but he didn't want to use his own ill-gotten money. In the time-honored American tradition, he asked the government to foot the bill. Fortunately, unlike today, where bailout and stimulus money is dispensed by the shovel-full to incompetent businesses, the government sent inspectors to make sure Redheffer's machine was a good investment.

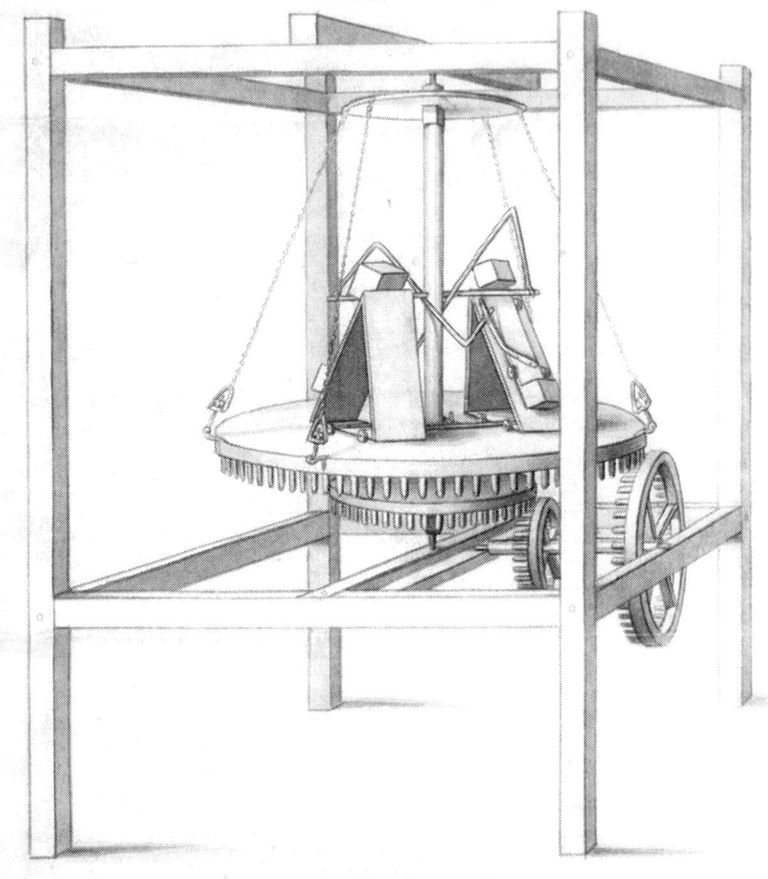

Redheffer's sketch of his alleged perpetual motion machine.

In January of 1813, eight inspectors examined the machine, but Redheffer made sure they didn't get too close—under the pretext that he didn't want anyone to break anything. It certainly looked like a perpetual motion machine, and Redheffer may have gotten his money, had it not been for the sharp eyes of the son of one of the inspectors.

There was a smaller machine connected to the alleged perpetual motion device by several gears, which Redheffer claimed was being powered by the larger machine. However, the son noticed that the wear pattern on the cogs of the gears indicated just the opposite—the small device *was actually powering* the large one!

It was clearly a case of fraud, but not being allowed to actually touch and examine the parts of the machine, the inspectors had to resort to a different approach. They commissioned Isaiah Lukens, an engineer, to build a similar machine, powered in the same manner, and they showed it to Redheffer. Realizing he had been discovered, he beat a hasty retreat to New York City.

Rather than learning from his experience and using his skills to build things that were actually useful, he chose instead to build another fake perpetual motion machine. Once again, the public was amazed and spent a lot of money to see the machine in action. Then one day there was a very special visitor—none other than Robert Fulton, the inventor who developed the steamboat.

Fulton noticed an irregular motion to the machine that made it wobble, and other things that sent up red flags. He accused Redheffer of fraud, and made the following challenge—Fulton would be allowed a close examination of the device, and if no fraud was uncovered he would pay for any damage he caused. Redheffer accepted the challenge, either through arrogance, overconfidence, or fear. Fulton then proceeded in an odd fashion—rather than taking apart the machine, he started taking apart the wall next to it.

After pulling down some boards, Fulton found a cord that went from the machine up to the floor above. When Fulton ascended the stairs to the next floor he found an old man eating a piece of bread with one hand—while with the other hand he was turning a crank to power the machine!

The visitors present at the time, angered by the discovery that they had been duped, smashed Redheffer's machine into little pieces. Redheffer quickly displayed his own motion—out the door to escape the incensed mob, and out of the city. Remarkably, he then returned to Philadelphia and was actually granted a patent for his fraudulent device in 1820! However, when he was still unable to get government funding, he and his machine disappeared into obscurity.

It is interesting to note that in the patent application, Redheffer designated the device as "power, machinery for the purpose of gaining." Unfortunately, apart from the admission fees, the only thing it really gained for Redheffer was notoriety in the annals of Bad Scientists.

Was it Dominant or Recessive Fraud?

What high school biology student hasn't heard of Gregor Mendel and his groundbreaking work with pea plants? This simple 19th century Augustinian monk and teacher was fascinated by the natural world and wanted to know how traits were inherited from one generation to the next. In a brilliant and detailed series of experiments that took many years, Mendel crossed pea plants and unlocked the secret of heredity. His paper, *Experiments with Plant Hybrids,* must rank among the most important scientific works of history, and would eventually create the foundation for modern genetics. (And just to prove how important it was, the established scientific community took 34 years to finally appreciate it.)

Mendel knew that one day his work would be understood and acknowledged, but he did a curious thing for a meticulous and dedicated scientist—he ordered that all his notebooks be burned after his death. That's something akin to an artist destroying his life's work. Mendel's excuse was that he didn't want future scientists to misinterpret his data, but perhaps there was a darker reason?

Gregor Mendel

In 1911, R.A. Fisher published a controversial paper suggesting that there were "statistical irregularities" to Mendel's results. In other words, they were just too good to be true. Fisher continued to analyze Mendel's existing data for the next twenty-five years, and then conclusively stated that many of Mendel's results were nothing short of fraud. Outraged defenders of the father of heredity sought to disprove this nasty accusation, but reluctantly, they could not.

Part of Fisher's claim was that Mendel repeated blocks of data to make it look like he had conducted more experiments, as well as changed the results of several experiments to make them appear more favorable, i.e., fit his predictions. To put it in terms of numbers, the probability of obtaining the near-perfect results of one set of experiments was one in 2,000, and in another, a whopping one in 33,000. Anyone who has ever conducted scientific experiments knows that Murphy's Law actively seeks victims wearing lab coats, so these numbers are quite disturbing. Either Mendel was the luckiest damn scientist who ever lived, or there was some hanky panky among the pea plants!

Arguments continue to rage on both sides, ranging from condemning Mendel for sullying the pure name of science, to excusing the irregularities as innocent mistakes (the latter arguments most likely devised by Enron attorneys). However, numbers make the loudest arguments, and for all of Mendel's good intentions (remember that's what paves the road to hell), it does appear as if he did do some creative statistical analysis.

But wait…In Mendel's favor is the overwhelming fact that he was right about recessive and dominant traits, and the basic theories of heredity. One must also consider that he was trying to convince some rather thick-headed scientists, so he needed to put his best statistics forward. And how can you get angry at a mild-mannered Augustinian monk who devoted his life to the search for knowledge?

And so, although it is somewhat out of character for this author not to slam down the avenging hammer of Bad Science, I can only deliver to this monk a slap on the wrist for fabricating some data for one of the most revolutionary concepts in the history of science.

Perhaps my avenging genes are not as dominant as I thought…

Hereditary Nonsense

One of the world's most influential naturalists was born in France in 1744. His full name was Jean-Baptiste Pierre Antoine de Monet Chevalier de Lamarck, but we will just refer to him as Lamarck so the printer of this book doesn't run out of ink. Actually,

ink figures prominently in the life of one of Lamarck's followers, but let's not put the fraud before the facts.

Lamarck's father sent him to study with the Jesuits, but as soon as his father died, he rode off to join the army. The day after he arrived, he so distinguished himself in battle that he was immediately promoted. He seemed destined for a great military career, until a friend picked up Lamarck by his head just for fun. (They must have been desperate for amusement in the French army.) Supposedly, this innocent act of horseplay led to severe glandular disease, and forced him to abandon his career in the military. (So remember kids, don't try picking up your friends by their heads.)

Portrait of Lamarck in 1824, when he was "Old and Blind." Some may argue he was blind for many years.

Anyway, the army's loss was science's gain. Lamarck's intense interest in botany led to several important books, as well as the official title of Botanist to the King, which wasn't half bad for a former Jesuit-army officer who had once been suspended by his skull. Later in life, his attentions turned to zoology, and he did some groundbreaking work in the study and classification of invertebrates.

However, for all his legitimate work in the various fields of science, Lamarck has been criticized for letting his imagination run wild. Volumes of material he produced have been determined to be worthless, because they were all based on pure speculation, which led to erroneous conclusions. But the bulk of the criticism leveled at Lamarck involved his theory of the inheritance of acquired traits.

For example, what this meant was that giraffe necks grew longer, because *they wanted* to reach the higher leaves, and prey animals developed more slender legs so they could run faster from predators. It was a case of form following function, and Lamarck described it as follows:

"The production of a new organ in an animal body results from the supervention of a new want continuing to make itself felt, and a new movement which this want gives birth to and encourages."

The changes brought about by these "new wants" were then supposedly "transmitted to the new individuals, which proceed from those which have undergone those changes"—i.e., inheritance of acquired traits.

One scientist who disagreed, August Weisman, set about to disprove this theory by lopping off the tails of mice, generation after generation. When none of the subsequent mice were ever born without tails, he believed he had proved that Lamarck was wrong. Undaunted, Lamarck claimed that the experiment wasn't valid, because the mice *did not want* their tails to be cut off, so it was not a naturally acquired trait. (I bet you didn't have to tell the mice that!)

Another interesting anti-Lamarckian example was that of the Jewish practice of circumcision. For thousands of years, Jews have wanted to perform the practice, but wouldn't you know, generation after generation of Jewish boys are still born with those pesky little foreskins. Yet despite hundreds of generations of valid data, Lamarckian supporters saw this as just as invalid as the amputated mice tails.

Before continuing with the fascinating history of acquired traits, it should be mentioned that Lamarck was also a supporter of the theory of spontaneous generation. He felt that the application of electricity and heat on gelatinous bodies created a "tension" and "orgasme" that could give rise to structure and life. (Doesn't make much scientific sense, but it sure sounds like fun!)

The slow but steady advance of scientific knowledge eventually discredited Lamarckian theory, but there are two twentieth century proponents worth mentioning in this bizarre saga. The first is Austrian-born Paul Kammerer, who believed he could force frogs to acquire new mating characteristics (I really don't want to know all the details), and then have these characteristics passed on to subsequent generations.

It appeared as though he was having remarkable success, until Dr. G. Kingsley Noble, the Curator of Reptiles for the American Museum of Natural History, paid a visit to Kammerer's lab in 1926. One of the allegedly inherited characteristics was black and swollen "nuptial pads," but upon closer inspection of the frogs it was found that the pads were black and swollen because they had been injected with India ink!

Of course, Kammerer claimed he was innocent, and that the fraud had been perpetrated by a disgruntled lab assistant, but the damage was done. Dr. Kingsley published his findings in *Nature*, and Kammerer's unnatural frog sex days were over.

He was invited to Moscow, where his ideas were still popular, but just three months after the scandal erupted, Kammerer committed suicide by shooting himself. However, his legacy and the Lamarckian theory continued with Ukrainian-born Trofim Denisovich Lysenko, who was lauded by the Soviets as a brilliant scientist, but in fact, barely had a scientific thought in his head. What he *did* have was the backing of Stalin, which was all he needed to make a shambles of the scientific community.

For example, in 1927, Soviet newspapers declared that Lysenko had found a method of fertilizing crops without using fertilizer. As proof, he supposedly grew winter peas on once-barren fields of Azerbaijan, a feat that promised to put food in the mouths of every peasant and cow. Unfortunately, future crops there all failed, but that minor fact was subsequently overlooked by the press.

Trofim Lysenko, with a face only Stalin could love.

Lysenko seemed to have particular skill in creating pseudo-scientific mumbo jumbo to describe his methods, without actually defining or explaining his new terms. He also harshly criticized legitimate scientists who saw right through his fraudulent practices. With the help of Stalin, however, he was able to circumvent this problem by having hundreds of the Soviet Union's brightest scientists imprisoned and executed—which is one way to end contrary opinions. As a result of this homicidal practice, the country's legitimate genetic programs were eliminated, and the geneticists along with them.

It wasn't until the 1960s, that genuine scientists were allowed to deliver genuine criticism against Lysenko, and once it all began hitting the Soviet fan the press jumped on the bandwagon and exposed Lysenko as a dangerous charlatan. And he did not escape Soviet justice, either. When he was finally removed as the head of the Academy of Sciences, they also took away his private bathroom. That will teach a person to destroy the infrastructure of the scientific community, suppress new ideas for decades, and cause the death of hundreds of highly educated scientists! Imagine the humiliation Lysenko felt having to use the public restrooms.

And so the peculiar saga of Lamarckian theory comes to a fittingly ridiculous conclusion. What began with a talented army officer being lifted by his head, led to a new theory of heredity, hundreds of years of misinformation, mouse and frog abuse, suicide, and murder, culminating in the revoking of bathroom privileges.

And so many people think that science is boring…

Getting the Cold Shoulder

So you're having a lovely stroll in the countryside of Switzerland, when you come across an enormous granite boulder perched on a bed of limestone that seems completely out of place. How did it get there, you wonder? Then you stop for a picnic, and while munching on some Swiss cheese you see a huge mound of rocks and dirt that looks as though it had been pushed there by some massive plow. What force could have caused that, you ask?

Well, if it had been in the late 1700s or early 1800s, scholars would have told you that these ancient features were the result of the Biblical Flood and drifting icebergs. Perhaps you then glanced up into the mountains where the summer sun was shimmering on an icy glacier, and you wondered out loud, "Couldn't that glacier once have been down in this valley, pushing this mound ahead of it, and when it receded it left behind these huge boulders, not to mention cutting those grooves along the walls of the valley?"

"Nonsense!" would come the angry reply. "I already told you these are the results of the Biblical Flood. It's ridiculous to think that glaciers move. Now shut up, eat your cheese, and don't ask anymore stupid questions!"

While this little scene may not have actually played out in this manner, it certainly illustrates the mood and popular scientific opinion about glaciers in Switzerland during that period of time. However, not everyone believed in static glaciers—like the people who lived near glaciers for generations and saw them move. Of course, they had no scientific training, only years of observation, so they didn't count.

In 1827, Swiss geologist Franz Hugi built a small hut on top of a glacier, and carefully marked its position with stakes, as well as chiseling reference marks into adjacent rocks. As expected, the hut changed position from the reference points, which should have proved that glaciers do indeed move. Yet it would be another ten years before someone came along who was determined to make the world listen and understand the truth about these massive sheets of ice.

Louis Agassiz was born in Switzerland, and first made his mark on the scientific world with fish. In fact, he became the preeminent authority on both living and fossil fish. His interest in glaciers began as something of a hobby, as he, too, had believed in the flood explanation—until he started seeing the evidence with his own eyes.

Louis Agassiz

Ignace Venetz and Jean de Charpentier told Agassiz about their belief that Switzerland had once been covered by glaciers, and they finally persuaded the reluctant fish expert to take some field trips. Agassiz gazed in wonder at the massive boulders—known as erratics—which appeared to have been tossed about by giants. There were the mounds of rocks and dirt—called moraines—that were like signposts along the course that the glaciers had taken. There were grooves and gouges cut into the valley walls, where receding glaciers left their calling cards. It was all quite remarkable and astounding, but it all made sense. Agassiz was now a believer, and he felt he had to enlighten the world.

Well, it may come as no surprise that the world, especially the scientific world, isn't big on being enlightened. In 1837, Agassiz presented his case for glaciers at a meeting of the Swiss Society of Natural Sciences. A performance of ancient yodeling techniques might have been better received. Undaunted, well, slightly daunted but still determined, Agassiz continued his work and traveled enough to come to the even more startling realization that most of Europe had once been covered by glaciers, which didn't make his theory any easier to swallow.

In fairness, it wasn't just his theory. For example, his good friend and colleague, Karl Schimper, had come up with the term *Eiszeit*, meaning Ice Age. Unfortunately, the two men squabbled over who should get credit for the glacier theory and ended up becoming enemies—which wouldn't be the last time in Agassiz's life something like that happened.

Despite the adversity, Agassiz persevered and published *Étude sur les glaciers* in 1840, and *Système glaciare* in 1847. Perhaps books on ancient yodeling techniques would have been better received.

Actually, slowly but surely (almost as slowly as glaciers move), some scientists were beginning to get it. But having been given the cold shoulder in Europe for so long, he was delighted to come to America and accept a position at Harvard. He was also delighted to travel around the Great Lakes region and see the same signs of former glaciers, which further expanded the icy grip of his Ice Age.

For all the credit Agassiz deserves, however, he was far from being infallible. After a trip to Brazil, he stated that this tropical country had been covered by glaciers as well, which it had not. He also held fast to the erroneous belief that the Ice Age had wiped out all life on Earth, and had created a clean slate for every plant and animal today—including mankind.

And don't even speak to Agassiz about Darwin and the theory of evolution, because he refused to believe that any force but God could create new forms of life. So adamant was Agassiz against evolution, and a few other things, that some of his students at Harvard jokingly formed the *Society for the Protection of American Students From Foreign Professors*. Agassiz wasn't laughing, and he had the students removed.

These little *faux pas* aside, Agassiz had lifted up the established field of geology and dumped it on its ear. He initiated a new science and changed the way the world looked upon its history. Had he lived for just a few more years (he died in 1873), he would have found that the tide of popular opinion had turned in favor of the Ice Age.

In his later years, Agassiz wrote that his only true accomplishment had been in his work with fish. Fortunately, he was wrong in that account, too, and will forever be recognized as the champion of the Ice Age.

Taking Out the Garbage?

In *An Essay on Criticism* written by Alexander Pope in 1709, he stated that a little learning was a dangerous thing. How right he was, especially when it applies to scientists who can affect the health of generations of people. As technology places ever-increasing power in the hands of modern scientists, they must be ever vigilant that they do not act too rashly with that "little learning" mankind has acquired.

Take the case of "junk DNA," a term coined in 1972 by Susumu Ohno. While most people should at least have some inkling that DNA (deoxyribonucleic acid) holds the body's "blueprints," few would be capable of accurately describing the intricacies of genes, mRNA, protein coding, base pairs, etc. They shouldn't feel too bad, however, as until relatively recently, most scientists believed that the majority of DNA was useless garbage. Even as late as 1980, Francis Crick (co-discoverer of the structure of the DNA molecule in 1953) referred to this alleged junk DNA as having "little specificity and conveys little or no selective advantage to the organism."

About only 3% of DNA is known as coding DNA—the active part that produces the building blocks of life. That leaves the other 97% of non-coding DNA, also known as introns (intragenetic regions), that didn't appear to do anything. Since scientists didn't see any purpose for all these introns, they labeled them as junk DNA. They were also disparagingly referred to as "excess baggage," useless "padding," and "selfish DNA" (as it appeared that their only purpose was to replicate themselves, much like many vacuous

Hollywood stars). In fact, when the project to map the entire human genome was proposed, some researchers even argued that they shouldn't bother mapping the junk portions, as it would only be waste of time.

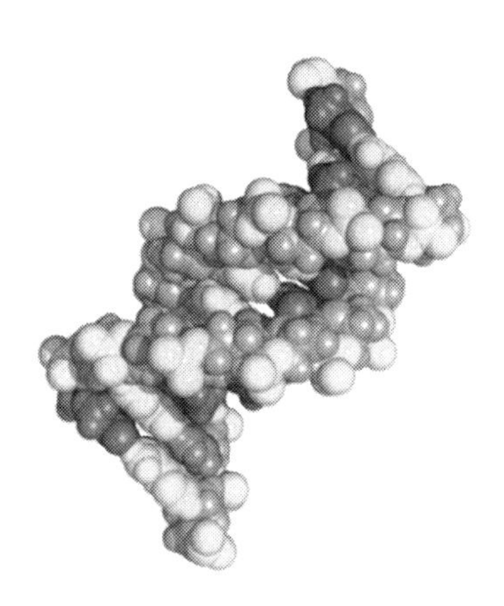
A section of DNA.

Fortunately, molecular biologists didn't initiate a program for "taking out the garbage," as in "cleaning up" human DNA by removing all that useless junk.

That *junk* that just happened to *constitute a whopping 97% of the DNA molecule...*

Didn't anyone stop to think that if something so critical and basic as DNA contained an overwhelming majority of something known as introns, that these introns might possibly have some purpose?

And didn't it even evoke the slightest suspicion when it was found that the more complex the organism, the more introns its DNA contained?

Just because you don't understand the purpose of something, it doesn't mean it has no function. Hell, if that's the case, since we don't know what 90% of the human brain does, why don't we remove all that useless gray matter?

And, as no one really sees a useful purpose for most politicians, maybe we should just get rid of all of them, too. (Wait, we may actually be on to something here...)

In any event, genetic light slowly began to dawn when researchers discovered that mice, rats, and humans had large pieces of junk DNA that were identical. If the stuff was so useless, some began to reason, why would it remain intact over 75 million years of divergent evolution among different species? Could it be that it wasn't all garbage?

To make sure these common introns were not just a fluke, they decided to see if these rodents and humans shared any introns with sequences of over 200 base pairs—in other words, sections so large that statistically they would be way beyond mere coincidence. To their astonishment, researchers found one. Or more accurately, they

 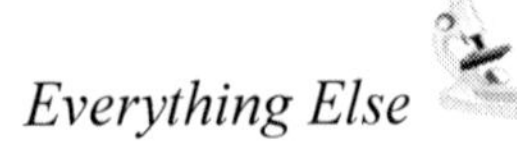

found *over 480 identical*, large introns shared by mice, rats, and humans!

Hmmmm, they thought. Perhaps if the junk DNA of very different species was carefully replicated and preserved over millions and millions of years, it did have a use after all. Suddenly, this discovery was like the brilliant light of Rudolph's nose finally dispelling the dark fog of genetic ignorance. Maybe a red-nosed reindeer and all those misfit introns had a purpose in life, too?

Which isn't to say we now understand all of this mysterious 97% of DNA, but that's okay. At least now we admit what we don't know, and research is currently underway to find out what introns do. So far, it appears as if sections of this not-really-junk DNA may regulate activity of the coding DNA, play a role during embryonic development, and possibly fix critical errors.

Where all this will ultimately lead is uncertain, but it could very well redefine the origins and development of life. Some have theorized that introns actually contain a complex genetic language just waiting to be translated. Perhaps in the near future, some Champollion of molecular biology will discover the genetic Rosetta Stone amongst the introns and decode nothing less than the meaning of life. And if we might speculate for a moment, what would be that first bit of wisdom that was revealed?

Could it possibly be, "A little learning is a dangerous thing…"

> "If the double helix was so important, how come you didn't work on it?"
>
> Ava Pauling to her husband, Linus, when the Nobel Prize was awarded to Crick, Watson, and Wilkins in 1962. Linus Pauling had already won the Nobel Prize for Chemistry in 1953, and also won the Nobel Peace Prize in 1962.

The Unkindest Cut of All

In Romania in 2004, 34-year-old Nelu Radonescu finally scheduled surgery to correct a testicular abnormality. Wanting the best surgeon available for the delicate procedure, he chose 56-year-old Dr. Naum Ciomu. It appeared to be an excellent choice, as

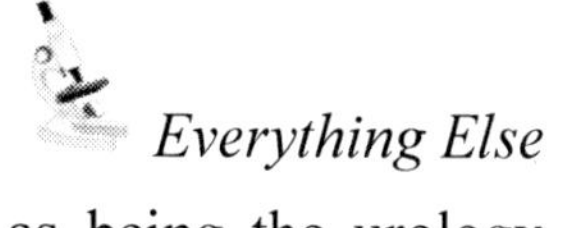

Ciomu was a professor of anatomy, as well as being the urology hospital's senior surgeon with years of experience.

Unfortunately, Dr. Ciomu was having a bad day, as he was admittedly "under stress" due to "personal problems."

(We interrupt this story to deliver a public service announcement: Gentlemen, if a surgeon is going anywhere near your genitals with a scalpel, make sure he is in a really good mood! And if you start to feel faint during the next paragraph, lay down with your feet elevated and skip to the next story.)

During the surgery, he accidentally cut Radonescu's urinary channel. Rather than correct his mistake and continue with the procedure, Dr. Ciomu flew into a rage, grabbed a scalpel and sliced off Radonescu's penis. To add insult to this monstrous injury, he then cut the severed penis "into three pieces before storming out of the operating theater."

Radonescu was rushed to another hospital where a plastic surgeon took tissue from his arm to rebuild something resembling a penis, and was at least able to restore the poor man's urinary function. Obviously, Radonescu—and his enraged wife—filed a lawsuit. The case was clear cut, so to speak, and they were granted $40,000 for penis reconstruction surgery. Additionally, Dr. Ciomu was ordered to personally pay $200,000.

That amount seems a mere pittance compared to the willful destruction of a man's penis, but even that amount brought outspoken criticism from Romanian doctors. By making Ciomu pay the settlement out of his own pocket, the doctors felt this case would "set a dangerous precedent" and that "doctors would in future avoid any cases where they could end up in court having to pay damages."

Hello, reality check here! What part of intentionally cutting off a man's penis and then chopping it into pieces because you're in a bad mood don't these doctors get? This was not some honest mistake, and Dr. Ciomu should be grateful Romanian courts don't impose the old "eye for an eye" punishment.

And what of the poor victim here, Nelu Radonescu? Sadly, his own words say it all.

"It will never be the same, but if I am even a quarter of

the man I was, I will still be very content."

It is doubtful that his wife will ever be content.

Breathtaking Inanity

Personally, there isn't anything that enrages me more than being falsely accused of something I didn't do, or being subjected to a similar form of injustice. However, acts of Bad Science are a very close second in their ability to make my blood pressure rise, as you may have repeatedly noticed throughout the pages of this book.

In this light, I have saved for last one of the most egregious and persistent acts of Bad Science that has ever slithered across the hearts and minds of humanity, with its grasping tentacles engulfing, enslaving, and blinding its ignorant and uneducated victims. (I know, why don't I just come out and say how I *really* feel about it.)

The war against the Theory of Evolution might be considered laughable if battles over it were not still being waged in the 21st century. With all the knowledge that modern science has provided us as to the age of the Earth, the long and detailed fossil record of life on this planet, and the genetic mechanisms that demonstrate how change occurs, it should all be about as close to a scientific slam dunk as you can get.

Yet remarkably, unbelievably, incredibly, there are *way* too many people who still think that the Earth is only about 10,000 years old, and life just suddenly popped into existence by a divine act. First presented as Creationism (actually an old belief, renamed to combat evolution), it has been repackaged and relabeled again as "Intelligent Design" (there's a misnomer if I ever saw one), and there are those today, who are still trying to force feed it down the throats of America's children.

But let me pause for a moment to calm down and get to the beginning of this story…

"As far as I can judge of myself I worked to the utmost during the voyage from the mere pleasure of investigation, and from my strong desire to add a few facts to the great mass of facts in natural science." -- Charles Darwin

This was how the shy and unassuming naturalist described his five years aboard the British ship, *Beagle*, as it sailed to South America, the Galapagos Islands, Tahiti, and Australia from 1831 to 1836. Those "few facts" he hoped to gather on the voyage turned out to be a treasure-trove of specimens and observations that would set Darwin's mind on a subsequent twenty-year journey that would culminate in 1859 in the publication of one of the most important books in history, *Origin of the Species by Means of Natural Selection, or the Preservation of Favoured Races in the Struggle for Life*. It was in this book that Darwin set forth the Theory of Evolution, and the best and worst of science and human nature was then set on a collision course. And it wouldn't take long for it all to hit the fan…

Charles Darwin in 1869.

There was no indication in Darwin's early years that he would become one of the most admired and reviled figures of his age. Born in 1809, it appeared as though he would follow in his father's footsteps and become a physician. However, shying away from the medical profession which appeared to be somewhat barbaric (which it was in those days, and probably still is, now that I think of it), he became interested in marine biology. His father then sent him to Cambridge to study to become a clergyman—as several of the clergy in England were also naturalists, this seemed to be a good way to earn a living from the church, and in your free time, poke around the countryside examining the flora, fauna, and geology.

Charles Darwin, however, had other ideas about joining the clergy, and when the opportunity came to sail around the world, it was like inviting a kid into a candy store. If fate had ever placed the right man in the right place, it was Darwin on the *Beagle*. His astute powers of observation and sharp mind saw deep below the surface

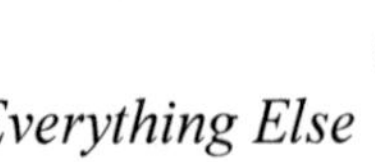

of the numerous fossils, bizarre plants and animals, and remarkable cultures he encountered. While he was not the first person to conceive of evolution, he was exposed to enough evidence to create a detailed and plausible model for it. (He was apparently also exposed to some disease on the voyage which severely affected his health the rest of his life, making the price he had to pay for his work even higher.)

But Darwin was hesitant to share his theory with the world, as he knew the scientific community was still dominated by religion. Even his own fiancé was afraid that her future husband would "burn" for his heretical belief that life on Earth had evolved from earlier forms that were millions of years old. It would be twenty years before Darwin could finally be persuaded to publish his findings in *Origin of the Species*, but it only took months before the critical backlash began.

One of the many satirical cartoons to which Darwin was subjected.

In June of 1860, at the annual meeting of the British Association for the Advancement of Science held at Oxford University, the evolutionists and creationists conducted a famous verbal duel. The highlight was undoubtedly the legendary response of pro-Darwin Thomas Huxley to Bishop Samuel Wilberforce's mischaracterization that Darwin believed that mankind was descended from the apes. Huxley allegedly stated something to the effect that he would personally rather be descended from an ape than from a bishop!

The heated debate devolved into all-out war that raged decade after decade. No matter how much new evidence came to light supporting Darwin, the creationists always managed to sidestep the facts. As the years rolled into a new century, the battles continued and the opposing sides were destined to meet on a stage even greater than that in the hallowed halls of Oxford. They were to meet in a small town in Tennessee in 1925, under the glaring light of the American press.

Thomas Scopes was a high school teacher who told his students about Darwin and the theory of evolution. The problem was, it was *illegal* to teach evolution in Tennessee, as well as in Arkansas and Mississippi. Yes, remarkable but true, it was actually against the law to teach a fundamental concept of science in the year 1925! (Of course, when one considers that women had only recently earned the right to vote, it reinforces the fact that American society was painfully primitive. And now that I think of it, it probably still is…)

The case went to trial, and like Wilberforce and Huxley sixty-five years earlier, two titans were about to clash. The fanatically religious William Jennings Bryan would lock horns with the shrewd and cunning defense attorney Clarence Darrow (who, in arguably the most amazing aspect of the trial, actually provided his services for free!). But Darrow and Scopes had two strikes against them from the start, as the very religious judge *denied the defense the right to call any expert witnesses*!

Darrow finally called Bryan, himself, to the stand, where the self-righteous prosecutor proceeded to make a fool of himself with his closed-minded and ignorant pronouncements. Then, in order to stop the trial on a winning note, Darrow pleaded guilty on behalf of his client. A fine of $100 was imposed upon Scopes, but not one penny had to be paid as the Tennessee Supreme Court eventually threw out the case. It had been an embarrassing national mess for the creationists, but it did not actually become legal to teach evolution in Tennessee until 1967!

But hold on to your evolved eohippus, the story does not end there, and indeed stretches into the new century. In 2004, in the Dover Area School District in Pennsylvania, the school board voted to make Intelligent Design part of the science curriculum. What that meant was, that the children would be taught that plants and animals were so complex that they could not have evolved on their own over the course of millions of years, that it had to be some "higher force" that created them. In other words, Intelligent Design basically says that because science is too complex for us to understand, let's fall back on religion. It's sooo much easier not to have to think…

Outraged parents cried foul and the case went to court. In December of 2005, Judge John Jones strongly ruled against Intelligent Design, stating it was clearly a thinly-veiled attempt at imposing religious beliefs, and it violated the separation of church and state. But he didn't stop there. In one of this author's favorite quotable quotes, the judge further characterized the nature of the school board's decision as being one of "breathtaking inanity."

Case closed.

Bravo!

Yet, sadly, the fight still continues as other school boards consider attempting the same maneuver. And there are still gullible audiences for numerous publications, cable television shows, and websites that practice the most unnatural selection—picking and choosing assorted "facts" that "prove" Creationism. I would love to think that all of this nonsense will fade and become extinct by the next century, but I have witnessed too much human ignorance to hold out much hope.

At least we have reached an age where scientists are no longer burned at the stake, but false ideas can still be very dangerous. If only religion would get the hell out of science and stay out—along with intolerance, jealousy, fraud, greed, and ego trips.

But alas, I must be dreaming. But what a fine dream it is, to pursue science for the sake of knowledge that can benefit all mankind and finally slay the dragon of ignorance!

Yeah, right.

I had better start putting together the material for the next volume of Bad Science…

> Our descendants will marvel at our ignorance.
>
> Seneca, c. 1 BC-65 AD

CPSIA information can be obtained at www.ICGtesting.com
Printed in the USA
LVOW11s0203151115

462622LV00001B/30/P

9 780979 900242